T0134957

Smart Innovation, Systems and Technologies

Volume 67

Series editors

Robert James Howlett, Bournemouth University and KES International,
Shoreham-by-sea, UK
e-mail: rjhowlett@kesinternational.org

Lakhmi C. Jain, University of Canberra, Canberra, Australia;
Bournemouth University, UK;
KES International, UK
e-mail: jainlc2002@yahoo.co.uk; Lakhmi.Jain@canberra.edu.au

About this Series

The Smart Innovation, Systems and Technologies book series encompasses the topics of knowledge, intelligence, innovation and sustainability. The aim of the series is to make available a platform for the publication of books on all aspects of single and multi-disciplinary research on these themes in order to make the latest results available in a readily-accessible form. Volumes on interdisciplinary research combining two or more of these areas is particularly sought.

The series covers systems and paradigms that employ knowledge and intelligence in a broad sense. Its scope is systems having embedded knowledge and intelligence, which may be applied to the solution of world problems in industry, the environment and the community. It also focusses on the knowledge-transfer methodologies and innovation strategies employed to make this happen effectively. The combination of intelligent systems tools and a broad range of applications introduces a need for a synergy of disciplines from science, technology, business and the humanities. The series will include conference proceedings, edited collections, monographs, handbooks, reference books, and other relevant types of book in areas of science and technology where smart systems and technologies can offer innovative solutions.

High quality content is an essential feature for all book proposals accepted for the series. It is expected that editors of all accepted volumes will ensure that contributions are subjected to an appropriate level of reviewing process and adhere to KES quality principles.

More information about this series at http://www.springer.com/series/8767

Foreword

For over a decade the mission of KES International has been to provide a professional community, networking and publication opportunities for all those who work in hi-tech knowledge-intensive subjects. KES focusses on the dissemination, transfer, sharing and brokerage of knowledge.

The KES community consists of several thousand research scientists, academics, engineers and practitioners who participate in KES activities. The community works in a range of disciplines including Intelligent Systems, Sustainable in Energy and Buildings, Sustainable Design and Manufacturing, Innovation and Knowledge Transfer and Digital Media.

Increasingly, both sustainable building and sustainable energy technologies benefit from the use of computer systems executing sophisticated algorithms that replicate some of the reasoning, problem-solving and pattern-matching abilities of the human brain—known as 'smart systems'. This amalgam of disciplines forms a vibrant theme for research that can be termed 'smart sustainability'.

Since 2009 KES has organized an annual conference on Sustainability in Energy and Buildings (SEB). The SEB conference attracts research not only on sustainable buildings and sustainable energy systems that employ conventional methods, but also on smart sustainable buildings and smart sustainable energy systems incorporating intelligent algorithms.

The first two conferences in the SEB series (SEB-2009 and SEB-2010) took place in Brighton, UK, under the auspices of the University of Brighton. SEB-2011 took place in Marseille, France, hosted by the LSIS laboratory. SEB-2012 took place in Stockholm, Sweden hosted by KTH Royal Institute of Technology. In 2013 SEB took the form of the Mediterranean Green Energy Forum linked with the World Renewable Energy Network and held in Fes, Morocco. SEB-2014 was organized by KES in partnership with Cardiff Metropolitan University, Wales, UK. In 2015 the conference went to Lisbon, Portugal under the UNINOVA research institute. The most recent iteration of the conference, SEB-2016, took place in Turin, Italy with Politecnico Torino.

This book *Smart Energy Control Systems for Sustainable Buildings* contains 11 chapters based on recent papers published in the SEB conference series on smart sustainability topics, extended and enhanced into book chapters. It forms an overview of leading-edge developments in this area.

KES International, UK   Professor Bob Howlett

Contents

Introduction

There is a widespread interest in the way that smart energy control systems, such as assessment and monitoring techniques for low carbon, nearly zero energy and net positive buildings at the micro level through to the neighbourhood scale to cities and beyond and how they can contribute to a Sustainable future, for current and future generations.

There is a turning point on the horizon as the supply of energy from finite resources such as natural gas and oil becomes less reliable, more volatile in economic terms and more challenging technically for extraction and more unacceptable socially, such as adverse public reaction to 'fracking'. Thus, these finite resources in 2017 are becoming more expensive, which is having a major influence on the design, optimisation, performance measurements, operation and preservation of: buildings, neighbourhoods, cities, regions, countries and continents. The source and nature of energy, the security of supply and the equity of distribution, the environmental impact of its supply and utilisation, are all crucial matters to be addressed by suppliers, consumers, governments, industry, academia, and financial institutions.

This book entitled *Smart Energy Control Systems for Sustainable Buildings* contains eleven chapters. Each of the chapters have been written by international experts, based on expanded and enhanced conference papers that were presented at the Sustainability and Energy in Buildings International conference series. The SEB International Conference has been hosted around Europe since 2009 organized by KES International in collaboration with University partners. Two of the largest and most successful SEB International Conferences took place in 2014 and 2016. In 2014, for SEB14, KES International collaborated with the Ecological Built Environment Research and Enterprise group in Cardiff School of Art & Design at Cardiff Metropolitan University, UK. In 2016, for SEB16, KES International collaborated with the Department of Energy at the Politecnico di Torino, in Turin, Italy.

The SEB International conference series has a number of General Tracks and Invited Tracks each year. A broad range of sustainability and energy-related topics relevant to the main theme of sustainability in energy and buildings, attract paper contributions each year. Since its formation in 2013, the General Track 'Sustainable Buildings' created and Chaired by Dr. John Littlewood at Cardiff Metropolitan University, UK has attracted the largest number of high-quality papers of any conference track since its formation in 2013. Applicable themes at SEB International Conferences include: sustainable design, construction and operation of existing and new buildings; neighbourhoods and cities (built and natural environment); modelling, monitoring and optimisation techniques; smart energy systems for smart cities; green information communications technology; and well as a broad range of solar, wind, wave and other renewable energy topics. The SEB International conference series brings together researchers and government and industry professionals to discuss the future of minimising energy use in the operation of buildings, neighbourhoods and cities from a theoretical, practical, implementation and simulation perspective.

This book includes new findings on the assessment, design, construction and optimisation of smart buildings and energy control systems for a sustainable future.

Topics include:—Control and optimisation of renewable energy systems (PV, wind energy, geothermal);—Highly efficient and holistic buildings and services for a sustainable future;—Methodologies & techniques for monitoring, evaluating and reporting on sustainable design, construction and services which affect building habitants': energy use and costs, comfort, behaviour and carbon emissions;—Smart monitoring and controls using intelligent systems;—Scenarios for a sustainable future and the role of energy storage. This book will provide a first class summary of the latest research in this area.

Summaries of the 11 chapters follow.

Chapter 1 authored by Salvatore Carlucci, Francesco Causone, Lorenzo Pagliano, and Marco Pietrobon discusses the challenges to reduce the existing gap between design and as-built building performance, suggesting that experimental data is required to characterize buildings. Living laboratories are discussed as one method of testing the effectiveness of different technologies to reduce the energy consumption or to increase the indoor environmental quality of the building as a whole; particularly for nearly zero-energy and the Passivhaus type buildings. The municipality of Mascalucia (Catania) in the Italian region of Sicily is discussed as the case study. The design process for the living laboratory approach was led by an optimization procedure based on extensive energy simulations; and resulted in an optimal energy balance and favourable thermal comfort conditions along the year. The detailed monitoring data is discussed following the commissioning phase, providing a first insight of the energy and comfort performance of the building.

Chapter 2 authored by Antonio Gagliano, Maurizio Detommaso, and Francesco Nocera1 discusses how green roofs can mitigate the urban microclimate, by reducing the surface temperature and hence the surrounding air temperature. The authors draw attention to the fact that the majority of existing Italy building stock was built without or with poor thermal insulation and require interventions for improving energy efficiency. Roofs are the component of the building envelope that are mainly engaged in both solar gains and thermal fluxes, and thus any refurbishment intervention should involve the design process. The chapter focuses on the analyses of the energy performance and the dynamic thermal behaviour of a green roof used in the retrofitting of an existing house situated in the southern part of Sicily in Italy. The results demonstrate the green roof refurbishment reduces the energy needs by 90% and 34% for the cooling and the heating periods, respectively; delays and attenuates the outdoor heat wave better of traditional roofs diminishing the average daily temperature fluctuations.

Chapter 3 authored by Magdalena Baborska-Narozny, Fionn Stevenson, and Paul Chatterton discusses the effectiveness of the collective learning that takes place in modern housing developments and how they can play a major role in terms of housing performance. Building performance evaluation (BPE) as a process for monitoring buildings through a series of stages is discussed and that it currently does not address the type and quality of collective learning processes happening within a community in relation to occupants using their new homes. The authors propose a Social Learning Tool to extend the BPE methodology and provide a framework to help building consultants and researchers understand the nature and degree of home user collective learning and community involvement better, which can in turn enhance the BPE process. A first partial application of the tool to six case study dwellings within a low carbon development in Leeds, within the north of the UK is discussed and which allowed identification of barriers and opportunities for collective learning.

Chapter 4 authored by James Pittam, Paul D. O'Sullivan, and Garrett O'Sullivan discusses how developing representative archetypes using a bottom-up approach for stock modelling is an excellent tool for evaluating the overall performance of the building stock. In Ireland there is no detailed housing database for Local Authority housing that catalogues the housing stock according to geometric configuration and thermal characteristics for each typology. A methodology to catalogue LAH stock and build a detailed housing stock database, using a GIS web based mapping application Google Street View is discussed. Eighteen house typologies across 36 LAH developments for Cork City in the South of Ireland is used as a dataset for demonstration of the methodology. A total of 10,318 housing units are counted and information subdivided into end of terrace, mid terrace, semi-detached, terrace lengths, orientation and elevation. This database then provides the base line assessment for building a stock aggregation model; the stock aggregation approach is used as a method to evaluate the energy performance of the building stock, beginning with analysis of individual house types; referred to as a 'bottom up approach'. The approach can be extended nationally very effectively with databases now being constructed remotely rather than the challenges of physical mapping and surveying. Using a matrix-based linear system for weighting parameters allows a large number of computationally efficient transformations of house types and parameters into archetypes. There is also flexibility in determining the grouping algorithm depending on the nature of the LAH studied. Investment in retrofit is highly justified in this area with large potential for reducing CO_2 emissions, the number of fuel poverty sufferers and victims of seasonal mortality due to thermally inefficient homes. The study suggests the method applied has scalable potential and is modular in structure facilitating wider adaptation.

Chapter 5 authored by Charikleia Moschou and Catalina Spataru discusses how the UK Government's Green Deal policy and how it intends to promote the energy efficient refurbishment of existing dwellings. Five different dwelling types are discussed as case studies with proposed intervention packages including the use of insulation, glazing and renewable energy technologies to offer the maximum amount of carbon dioxide savings. A model has been developed in order to evaluate the recommended packages, which demonstrates that insulation to the internal cavity, floor, roof/loft, doors and micro turbine would offer the most significant CO_2 reductions in detached dwellings. Whilst insulation to the external walls, cavity, floor, roof/loft, doors and a wind turbine offer the lowest amount of CO_2 savings in terraced dwellings.

Chapter 6 authored by Rainer Elsland discusses how to evaluate the useful energy demand for space heating purposes, norm-based bottom-up models are applied that capture the energy-related characteristics of buildings. Since electricity demand per household attributed to appliances and lighting has increased substantially for EU citizens since the late 1990s, questions arise whether the static, norm-based approach is underestimating the contribution of internal heat gains to covering useful energy demand. To analyse the impact of dynamic internal heat gains in the residential sector, a bottom-up model is applied that covers the EU27 building stock with a country-specific typology. The study reveals that the norm-based approach underestimates the contribution of internal heat gains to covering thermal heat demand. Comparing the countries throughout the EU27 climate zones indicates that, on average, the static and the dynamic share of internal heat gains up to 2050 vary in a range of 20–70%.

Chapter 7 authored by Dong-Luong Dinh and Tae-Seong Ki discusses a novel hand gesture recognition system for appliance control in smart homes using the labelled hand parts via the trained RFs from a hand depth silhouette. The chapter discusses that dwellings are an important place for people especially the elderly and disabled and the home environment not only affects the quality of life, but it is also a place where people spend a large amount of their time. Applying advanced technologies in various fields of architectural, electrical, automatic control, computer, and biomedical engineering to home is getting a lot of attention from researchers to create smart home. One of the important technologies for smart home is how to control home environments. Potential applications of such a human interaction based on hand gesture recognition include home entertainments, home appliances control, and home healthcare systems; particularly in Vietnam where case study is drawn from for this chapter. The mean recognition rate of 98.50% over the four hand gestures from five subjects is discussed. The authors indicate that the proposed hand gesture recognition method should be useful in automation applications for appliance control in smart home environment.

Chapter 8 authored by Filipe Rodrigues, Carlos Cardeira, and J.M.F. Calado discusses the importance of understanding household daily consumption in order to design and size suitable renewable energy systems and energy storage. Artificial Neural Networks (ANN) are shown as the possibility to forecast the electricity consumption of a household with certainty, which are a recognized potential methodology for modelling hourly and daily energy consumption and load forecasting. Input variables such as apartment area, numbers of occupants, electrical appliance consumption and Boolean inputs as hourly meter system were considered. It is discussed that a feed-forward ANN and the Levenberg–Marquardt algorithm provided a good performance. The case studies for this work were 93 real households, in Lisbon, Portugal, between February 2000 and July 2001, including both weekdays and weekend. The results show that the ANN approach provides a reliable model for forecasting household electric energy consumption and load profile.

Chapter 9 authored by John Cosgrove, John Littlewood and Paul Wilgeroth discusses the background to energy usage in production operations and sets out some principles, process steps and methods to provide a more holistic view of the Significant Energy Users (SEUs) and the related consumption of energy and technical services (Heat, Air, Water). A model based on direct and indirect energy analysis from a 'product' viewpoint is extended to identify waste or auxiliary energy in line with 'Lean' principles. The proposed process mapping methodology (Value Stream Mapping (VSM)) effectively acquires production and energy data that can be modelled to provide both steady state and dynamic energy consumption and potentially provide a multidimensional hierarchical view of this energy consumption and cost directly related to production equipment. The method is one that can be updated easily to reflect changes in the production environment and to provide a holistic view of the energy and technical services in the context of the varying production activity

Chapter 10 authored by Jorn K. Gruber and Milan Prodanovic discusses how the energy sector has undergone an important transformation as a result of technological progress and socioeconomic development; with the continuous integration of renewable technologies drives the gradual transition from the traditional business model based on a reduced number of large power plants to a more decentralized energy production. The increasing energy demand and intermittent generation of renewable energy sources require modern control strategies to provide an uninterrupted service and guarantee high energy efficiency. Utilities and network operators permanently supervise production facilities and grids to compensate any mismatch between production and consumption. The enormous potential of local energy management contributes to grid stability and can be used to reduce the adverse effects of load variations and production fluctuations. A building energy management system which determines the optimal scheduling of all components of the local energy system is proposed with a two-stage optimization, based on a receding horizon strategy and minimizes two economic functions subject to the physical system constraints. The performance of the proposed building energy management is validated in simulations and the results are compared to the ones obtained with other energy management approaches.

Chapter 11 authored by Gráinne McGill, Tim Sharpe, Lukumon Oyedele, Greg Keeffe, and Keith McAllister discusses how the adoption of the German Passivhaus Standard in the UK has grown rapidly, which has been stimulated by the shift towards energy efficient design and rising fuel costs. The concept is perceived as a potential means of meeting energy and carbon targets through an established, reliable methodology. However, the performance of the Standard in terms of adequate indoor air quality and thermal comfort in a UK climate remains under-researched. This paper describes the use of the Passivhaus Standard in a UK context, and its potential implications on indoor environmental quality. A case study is presented, which included indoor air quality measurements, occupant diary, building survey and occupant interviews in a Passivhaus social housing project in Northern Ireland. The study found issues with indoor air quality, the use and maintenance of Mechanical Ventilation with Heat Recovery (MVHR) systems, lack of occupant knowledge and the perception of overheating in the case study dwellings. The findings provide a much needed insight into the indoor environmental quality in homes designed to the Passivhaus standard, which can be disseminated to aid the development of an effective sustainable building design that is both appropriate to localized climatic conditions and also sensitive to the health of building occupants.

Cardiff Metropolitan University, UK   Dr. John Littlewood

Chapter 1
Zero-Energy Living Lab

Salvatore Carlucci, Francesco Causone, Lorenzo Pagliano
and Marco Pietrobon

Abstract In order to reduce the existing gap between design and as-built building performance, experimental data are required, both for single components and for whole building characterization. Real scale buildings, designed and equipped to perform such as living laboratories, may allow testing the effectiveness of different technologies to reduce the energy consumption and/or to increase the indoor environmental quality of the building as a whole. In order to understand to what extent the nearly zero-energy and the Passivhaus concepts could be extended to the Mediterranean climate, a new building (a detached single family house) was designed and constructed in the municipality of Mascalucia (Catania), in the Italian region of Sicily. It has been conceived to perform such as a living laboratory able to provide useful information on the actual performance of the building and its components, such as the Air to Earth Heat Exchanger. The design process was led by an optimization procedure based on extensive energy simulations. It resulted in an optimal energy balance and favourable thermal comfort conditions along the year. The building is equipped with an accurate Building Automation and Control System (BACS), and a number of sensors for a detailed energy and environmental monitoring. The early monitored data, following the commissioning phase, provide a first insight of the energy and comfort performance of the building. Further results including improvements in control algorithms and a comprehensive data analysis are ongoing and their completion is expected in a short time.

Keywords Living lab · Zero-energy building · Passive building · Mediterranean climate · Energy simulation · Energy monitoring · Energy optimization

S. Carlucci
NTNU Norwegian University of Science and Technology,
Department of Civil and Environmental Engineering, Høgskoleringen 7A,
7491 Trondheim, Norway
e-mail: salvatore.carlucci@ntnu.no

F. Causone (✉) · L. Pagliano · M. Pietrobon
Politecnico di Milano, Energy Department, End-Use Efficiency Research Group,
Via Lambruschini 4, 20156 Milan, Italy
e-mail: francesco.causone@polimi.it

© Springer International Publishing AG 2017
J. Littlewood et al. (eds.), *Smart Energy Control Systems for Sustainable Buildings*,
Smart Innovation, Systems and Technologies 67,
DOI 10.1007/978-3-319-52076-6_1

1.1 Introduction

The European construction industry received a substantial boost towards improving buildings energy and environmental performance as consequence of the Energy Building Performance Directive [1] and its recast version [2]. A similar change has been experience also by the US construction industry, fostered by different drivers [3]. Nevertheless, researchers and professionals report that a gap in building energy and carbon performance still exist between design and as-built [4]. The major reasons of this performance gap are:

- Lack of information on the real dynamic performance of building components;
- Inadequate characterization of occupant behaviour;
- Building systems failure or inadequate maintenance and operation;
- Inconsistencies between design and construction;
- Lack of, or inadequate, post occupancy evaluation or commissioning;
- Limitations in the spatial density and quality of data contained in weather databases such as Test Reference Years (TRYs);
- Limitations in the simulation algorithms used by energy simulation software.

A defined and structured learning loop [4–6] is clearly still missing in the construction industry. This may be developed targeting some fundamental aspects:

- Improving cooperation between contractors, designers and researchers;
- Enhancing education levels (up to Ph.D.) of new professional actors in order to boost innovation and learning loops;
- Improving formation and information of operators (workers and professionals) at any level of the construction process (design, construction and commissioning);
- Establishing shared standard for the characterization of real dynamic properties of building components;
- Defining shared procedures for in situ measurements: test cell laboratories and real scale buildings (living labs);
- Defining shared procedures on modelling occupant behaviour in simulation tools;
- Improving energy simulation software and integrating energy simulation with design tools (a useful tool could be Building Information Management—BIM);
- Adopting shared weather data monitoring and acquisition procedures to define comparable and reliable TRYs and actual weather data files;
- Developing commonly accepted procedures for the preparation of future weather data files based on climate change models;
- Defining new quantitative post occupancy evaluation or commissioning procedures of the entire building;
- Managing building construction from the inception to the operation (a useful tool could be Building Information Management—BIM).

On-going research projects are trying to challenge some of the issues listed above, and a few innovative practices and contractors are contributing with private investments.

All of the above issues are important and vital to the final goal to reduce the gap between design and as-built performance. However, when talking about high performance buildings, zero-energy buildings or positive energy buildings, what is currently missing the most is the availability of actual operational data, showing to which extent such ambitious performance targets are actually reached and with which short and long term comfort levels. In case of misalignment between design aims and real performance, it is also important to identify the causes of such misalignment. Real scale buildings, designed and equipped to perform such as living laboratories, may allow testing various technologies and their integration in the energy concept as for their effectiveness in reducing energy consumption and/or increasing indoor environmental quality. Living labs may furthermore be used to obtain occupant behaviour models based on real data, and to check the influence of different building systems operation strategies on the final performance of the building. Whole scale buildings allow characterizing the global operational building performance.

By 31 December 2020 (31 December 2018 for buildings occupied and owned by public authorities), all new buildings in EU member States should be nearly zero-energy buildings. EU member States shall draw up national plans for increasing the number of nearly zero-energy buildings, reflecting national, regional or local conditions.

Many definitions of zero-energy building are available in the literature [7, 8], dealing with how to establish the energy balance (monthly, yearly etc.), what kind of energy or other indicator use in the balance (primary energy, exergy, equivalent CO_2 emission etc.), and on what boundaries to consider (the building walls, the construction site, the district, etc.).

The definition reported in the recast version of EPBD [2], although open to some interpretations, sets a few fundamental principles. A nearly zero-energy building means a building (Art. 2):

- That has a very high energy performance, i.e. it has a nearly zero or very low amount of energy needed to meet the energy demand [...], which includes, inter alia, energy used for heating, cooling, ventilation and lighting (art. 2);
- For which the nearly zero or very low amount of energy required should be covered to a very significant extent by energy from renewable sources, including energy from renewable sources produced on-site or nearby.

Moreover, according to Art. 9 of EPBD [2], Member States must include in their national plans, inter alia, the Member State's detailed application in practice of the definition of nearly zero-energy buildings, reflecting their national, regional or local conditions, and including a numerical indicator of primary energy use expressed in kWh/m^2 per year. The study promoted by the EU Commission in order to set a frame-work for those National Plans [9] concluded that a definition of nearly zero-energy building must be based on four elements:

- A performance part on energy needs and energy use (that is a part whose objective is to give a quantitative expression to the principle of reduction of energy demand as close as possible to end-use; this part would include indices as energy needs for heating, cooling and production of hot water and energy use for lighting (and optionally energy use for ventilation, auxiliaries and plug loads);
- A yearly weighted primary energy balance defined as in EN 15603 [10], that takes into account both the energy delivered by the grid(s) to the building and the energy generated at the building site and exported to the grid(s)—preferably based on monthly or shorter time intervals;
- The interaction of the building and the on-site generation from renewable energy sources (RES) with the grid, should be quantified by means of, for example, a 'load matching index' or other similar indices—in the end showing the share of self-consumed locally generated renewable electricity, calculated with time steps of a month, day or (preferably) hour, and the impact on the grid, for example to which amount the grid is used as a virtual inter-seasonal storage by the building (thus transferring costs from the building to the grid);
- One or more long-term comfort indices calculated according to EN 15251 or other relevant literature, because "an energy declaration without a comfort declaration makes no sense" [11]. IEA Annex 52 'Towards Net Zero Energy Solar Buildings' has analysed and proposed methodologies for incorporating comfort indexes in the characterisation of zero-energy buildings [12]. In any case, energy-related benchmarks for nearly zero buildings must include the underlying comfort level in an explicit and quantified manner.

A good reference to set benchmarks for the energy needs, relatively easy to meet by energy from renewable sources produced on-site, is the Passivhaus certification method originally developed for countries with a continental climate (e.g., Germany), where the major challenge is to contrast the low outdoor temperature and to wisely exploit the internal and solar heat gains, while providing adequate indoor air quality levels. Cooling and dehumidification are usually less important in this climate, especially for residential buildings. The technologies developed to comply with the Passivhaus certification method in the continental climate, are not necessary effective also in other climates. Probably they need to be adapted to the different climatic challenges and to be complemented by new or different technologies [13]. The PH certification has been extended to warm climates under the Passive-on Project [14] in order to explicitly include energy needs for cooling. It, in fact, requires energy need for space heating lower than 15 kWh/m^2 per year, energy need for cooling and dehumidification lower than 15 kWh/m^2 per year, primary energy for all domestic applications (heating, hot water and domestic electricity) lower than 120 kWh/m2 per year, and air tightness at 50 Pa (n_{50}) lower than 0.6 air change per hour (ACH).

In order to analyse and show to what extent the nearly zero-energy and the Passivhaus concepts may be extended to the Mediterranean climate, a new building (a detached single family house) was designed and constructed in the municipality of Mascalucia (Catania), in the Italian region of Sicily [15]. It has been conceived to perform such as a living laboratory able to provide useful information on the actual performance of a zero-energy building in the Mediterranean climate.

Energy simulations were used to support the development of an optimized detailed design, by using specified long term comfort indexes as objective functions of the optimization algorithm, as described in the following sections. The concept is based on an accurate control of building systems and components. The Building Automation and Control System (BACS) has been complemented with several sensors, which provide a detailed monitoring of energy flows and comfort conditions, useful for the management of the building and to further improve the design concept for future constructions.

1.2 The Climate Challenges

The Mediterranean climate—Csa/Csb under the Köppen climate classification [16–18]—is a particular variety of subtropical climate. The lands around the Mediterranean Sea form the largest area where this climate type is found [19]. The majority of the regions with Mediterranean climates have relatively mild winters and very warm summers. Because most regions with a Mediterranean climate are near large bodies of water, temperatures are generally moderate with a comparatively small range of temperatures between the winter low and summer high (although the daily range of temperature during the summer is large due to dry and clear conditions, except along the immediate coasts) [19]. Under the Köppen-Geiger system, "C" zones have an average temperature above 10 °C in their warmest months, and an average in the coldest between 18 and −3 °C. Areas with this climate receive almost all of their precipitation during their winter season, and may go anywhere from 4 to 6 months during the summer without having any significant precipitation [19].

Building cooling is the most challenging issue in the Mediterranean climate, due to the strong solar radiation and to the high ambient temperature. The large daily range of temperature during the summer provides, nevertheless, a considerable potential for nigh-time ventilative cooling [14].

The Mediterranean climate, as described under the Köppen-Geiger system [19], provides a high potential to exploit passive strategies for building design [14].

The use of thermal mass inside the building (mostly the floor slab) has been shown effective to reduce heating demand, when coupled with thermally insulated walls, because it provides a useful storage of internal and solar heat gains along the day [20]. This sDuring the cooling seatrategy showed to be mostly effective when thermal insulation is on the outer face of the wall [20]. Even from the thermal comfort point of view, internal thermal mass may provide a more homogeneous and acceptable surface temperature distribution and smooth and delay temperature changes along the day [21].

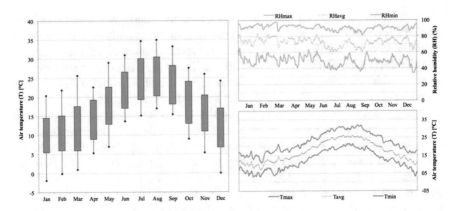

Fig. 1.1 Catania (Italy) weather data: (*left*) maximum and minimum values of the outdoor air temperature and interquartile range per each month (*blue bar*); (*right*) daily maximum, average and minimum outdoor air temperature and relative humidity

During the cooling season, the use of thermal mass inside the building may prove effective, if coupled with night-time ventilation, to reduce cooling loads and to enhance thermal comfort [20]. Due to the high daily range of temperature (Fig. 1.1), night-time ventilation may be easily exploited in the Mediterranean climate, especially if vegetation is growing in the surroundings of the building. An accurate control of solar gains is nevertheless required, because sufficiently ample charge/discharge loops of building thermal mass may not be provided by means of night-time ventilative cooling, for extreme heat gains. The orientation and size of the windows become therefore fundamental parameters to reduce heat gains, together with building shape [20], solar shading devices and glazing g-value [22].

An effective way to exploit thermal mass in low-rise buildings is an exposed floor slab, with some thermal contact with the ground. In this case the heating demand may slowly increase, but cooling loads may substantially decrease due to some coupling with the large thermal inertia of the ground. Night ventilative cooling also makes use of the interaction with thermal mass and particularly with the floor that has been partly exposed to solar radiation during the day [20]. The cooling effect of ventilation may be further extended during the day if ground thermal mass is exploited by means of an Earth to Air Heat Exchanger (EAHE) [23]. An accurate management of the mechanical or mixed-mode ventilation system is nevertheless required, and it may be guaranteed only by a BACS.

The use of thermal insulation layers in walls and roof has been proved useful also in the Mediterranean climate by several authors [20–22]. During the heating season, the position of the thermal insulation layer is not substantial, although it was shown that heating demand may be lower when thermal insulation is on the outside face of the wall [20]. This is true for passive buildings with no heating system, while in building with a heating system, performance substantially changes on the basis of the building system and its control. The position of the thermal insulation layer on the outdoor face is nevertheless suggested to decrease moisture condensation risk.

During the cooling season, the position of the thermal insulation layer is, instead, important, and external insulation was considered the most effective solution in the Mediterranean climate [20]. According to standard EN ISO 13786 [24] the dynamic thermal characteristics of the building envelope, which mostly affect cooling load are:

- The decrement factor: ratio of the modulus of the periodic thermal transmittance to the steady-state thermal transmittance (U-value);
- The time shift: period of time between the maximum amplitude of a cause (ambient temperature) and the maximum amplitude of its effect (internal heat flux);
- The periodic thermal transmittance: complex quantity defined as the complex amplitude of the density of heat flow rate through the surface of the component adjacent to zone m, divided by the complex amplitude of the temperature in zone n when the temperature in zone m is held constant;
- The thermal admittance: complex quantity defined as the complex amplitude of the density of heat flow rate through the surface of the component adjacent to zone m, divided by the complex amplitude of the temperature in the same zone when the temperature on the other side is held constant.

When the thermal insulation layer is moved to the external face of the wall, the periodic thermal transmittance decreases, and so it does the decrement factor, while the thermal admittance at the internal surface increases. This is coherent with the fact that there is hence active thermal mass directly exposed to internal air that is necessary for the effectiveness of summer night ventilative cooling and has a stabilising effect on the indoor operative temperature all-year round.

The use of thermal insulation on the external face of the wall is therefore suggested in the Mediterranean climate, and may be coupled with low solar absorbance and high emissivity finishing materials, which reduce the surface temperature of external surfaces.

Following the recommendations reported so far, and adopting a detailed energy simulation engine, such as EnergyPlus [25], and an optimization engine, such as Gen-Opt [26], it is possible to design a passive building according to the Passivhaus certification (originally developed to substantially decrease heating loads), characterized also by a very low cooling demand. Due to the very low total energy demand along the year, and to the large amount of available solar radiation, it is possible to reach a zero-energy balance, on yearly basis, by means of reasonably sized on-site photovoltaic and solar thermal systems.

A list of design hints obtained or confirmed by the optimization process for high performance buildings approaching a zero-energy target in the Mediterranean climate may be summarized as follow:

- Self-shading building (e.g., U-shaped);
- External vegetation and pervious paving materials;

- Controllable solar shading and high performance glazing;
- Low solar direct absorptance and high emissivity finishing materials on external wall and roof ('cool roof' and 'cool walls');
- High thermal insulation on the external face of walls and roof;
- Indoor thermal mass exposed (mostly floor and internal face of walls);
- Mixed mode ventilation and nigh-time ventilative cooling;
- Coupling of the building with the ground: low thermal insulation of the ground floor and EAHE;
- BACS for building systems optimization and monitoring;
- Photovoltaic and solar thermal panels i.e. on site renewable energy conversion and exportation.

1.3 The Building

The living lab was designed to understand how far the nearly zero-energy and the Passivhaus concepts may be extended to the Mediterranean climate. It is a single family detached house, located in the municipality of Mascalucia (Catania) in the Italian region of Sicily (Figs. 1.2 and 1.3). It has been designed targeting the requirements of the Passivhaus certification method:

- Energy need for space heating lower than 15 kWh/m^2 per year;
- Energy need for cooling and dehumidification lower than 15 kWh/m^2 per year;
- Primary energy for all domestic applications (heating, hot water and domestic electricity) lower than 120 kWh/m^2 per year;
- Air tightness at 50 Pa (n_{50}) lower than 0.6 air change per hour (ACH).

The high envelope performance (Table 1.1) is complemented by the local production of renewable energy by means of photovoltaic modules and a solar thermal system and by the exploitation of geothermal energy by means of an EAHE in the mechanical ventilation system. A thick external thermal insulation layer made of

Fig. 1.2 The building (*left*) in different construction phases: completion of mineral wool thermal insulation (*centre*) and structural concrete and masonry elements (*right*)

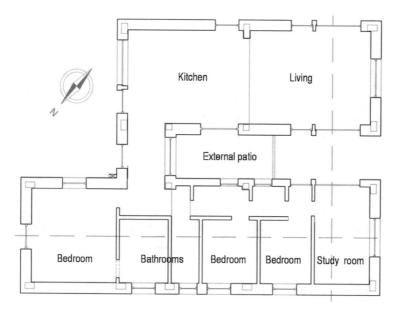

Fig. 1.3 Plan view of the building with orientation

mineral wool, triple glazing windows, frame profiles with thermal cut, and great care in construction details, are used to achieve high thermal insulation and air-tightness levels. Thermal mass exploitation is provided mainly by the concrete floor and roof slabs directly exposed to indoor air, and by masonry walls with external insulation. Natural cross ventilation is enhanced by windows disposition and by the external patio (Fig. 1.3).

In particular, the EAHE provides pre-heating or pre-cooling to the air supplied by the ventilation system. The supply-air temperature can be further adjusted by means of a heat recovery unit, with automatic bypass of the EAHE when temperatures allow it, and by a heating/cooling coil before entering the indoor environment. The solar thermal system is integrated with a heat pump generator. The complex system is automatically regulated by a BACS supported by Konnex (KNX) protocol. The dwelling also benefits from natural ventilation (cross ventilation), especially for night cooling.

The building is a living lab, so equipped with a wide variety of sensors inside the building, outside the building and in the different sections of the building systems, as described in detail in the following sections. It is furthermore equipped with an vapour compression heat pump, useful to provide both heating and cooling. This opportunity is considered useful to compare energy and indoor environmental performance under different operational conditions:

- Free-running, when the building is operated as a purely passive building without heating/cooling active systems;
- Heating;

Table 1.1 Building main features

Description or building feature	Value or range of values
Project name and location	*Progetto Botticelli* in Mascalucia (Sicily)
Building type	Detached single family house
Conditioned floor area	144 m^2
External walls transmittance	0.13 W/(m^{2}K)
Roof thermal transmittance	0.13 W/(m^{2}K)
Basement thermal transmittance	0.23 W/(m^{2}K)
Windows thermal transmittance	0.90–1.10 W/(m^{2}K)
Envelope air tightness (n_{50})	Lower than 0.60 h^{-1}
Construction type	Structural concrete and masonry, with mineral wool thermal insulation

- Cooling;
- Heating and cooling.

The comparison of data gathered during different operational conditions may result extremely useful for: designers, researchers, facility managers, energy managers, managers of district heating/cooling systems, managers of electrical grids, decision makers.

The building is moreover designed to be resilient to extreme weather condition, for example those experienced in summer 2003, and may be operated targeting also extremely challenging comfort conditions, excluding only humidity control, that is not usually an issue in residential buildings.

1.4   Simulation and Optimization of the Design Concept

Mascalucia is a small town on the slopes of Etna volcano. Although the construction site is less than 10 km far from the sea, the house is located at a height of 420 m above the sea level. Despite Mascalucia falls in the category Csa[1] according to the Köppen-Geiger classification system, as presented in Sect. 1.2, the local climate is characterized by a quite wide night to day temperature fluctuation, especially during summertime, that makes effective the adoption of night-time ventilation for cooling down building massive components, such as slabs and walls.

---

[1]Csa: Mild with dry, hot summer. Warmest month has average temperature higher than 22 °C. At least four months with average temperatures over 10 °C. Frost danger in winter. At least three times as much precipitation during wettest winter months as in the driest summer month (http://www.eoearth.org/view/article/162263/).

In order to exploit this climate opportunity, to tackle the challenges discussed in Sect. 1.2, and to identify optimal features of building envelope components for minimizing seasonal thermal discomfort, an energy model has been developed. The energy design of a building is nevertheless a multivariable problem, which can accept different sets of solutions. A mathematical optimization technique has been, therefore, used to guide the simulation engine.

Since detailed building performance simulation tools are increasingly used in the design of buildings and available computation power is rising, optimization may now be accomplished in a relatively short time and, hence, be compatible with the time scale of a building design.

1.4.1 Mathematical Optimization

Mathematical optimization (in the following just optimization), is an automated procedure that explores a very large number of variants, called *problem space*, in an accessible time domain, and looks for those variants which better perform with respect to one or more quantitative goals of an optimization problem, called *objective functions*. Optimization dealing with one objective function is called *single-objective optimization*; otherwise, it is called *multi-objective optimization*. However, two or more objective functions can be analytically combined in a single mathematical structure called *utility function*, and this process is called *scalarization*. During an optimization run, an algorithm guides a simulation engine to simulate those building configurations that conventionally minimize objective functions. Some constraints may be used to limit the problem space; in that case it is called *constrained optimization*, otherwise it is called *unconstrained optimization* [27].

Our analysis is intended to identify the features of the building envelope components and the control strategies of selected passive strategies, which minimize thermal discomfort by adopting a constrained scalarized optimization. In this work, thermal discomfort is assessed computing a long-term thermal discomfort index: the Long-term Percentage of Dissatisfied (LPD) [28], during both summer and winter. Summer and winter LPD (indicated respectively with LPD_s and LPD_w) are hence combined in a utility function $U(x)$, such that:

$$U(\mathbf{x}) = LPD_S^2(\mathbf{x}) + LPD_W^2(\mathbf{x}), \tag{1.1}$$

which graphically measures the square of the Euclidean distance of a given variant from the origin of the two-dimensional domain, constituted by the two objective functions. In Eq. (1.1), $\mathbf{x}$ is the vector of all design variables that characterizes a given building variant.

The options for the opaque envelope components have been selected combining three levels of the steady-state thermal transmittance (U) and three levels of the time shift (S). The three levels are denoted with the signs "+", "o", "-", in order to

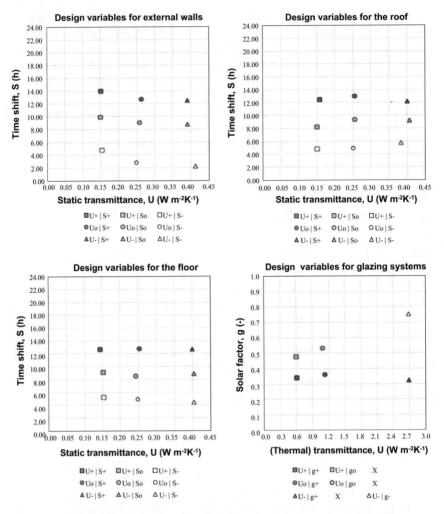

Fig. 1.4 Values of physical properties of tested design variables regarding external walls, roof, floor and glazing units

identify respectively a high-performance, a medium-performance and a low-performance option, regarding a specified physical parameter (Fig. 1.4). Furthermore, six types of glazing units characterized by different values of transmittance and solar factor are considered (Fig. 1.4). Because the daily distribution and the quantity of solar radiation are quite different for each of the façade orientations, the six types of glazing units are tested individually on each of those orientations: southeast, southwest, northeast and northwest. Moreover, due to the U-shape of the building, the six glazing solutions are also tested around the central court.

Table 1.2 Options for controlling solar shading devices, and window opening during summer nights

Design variable	Physical quantity	Code	Option
Control strategy of shading devices	Indoor dry-bulb air temperature	InTemp	$T_{air,int} > 25\ °C$
	Outdoor dry-bulb air temperature	OutTemp	$T_{air,out} > 25\ °C$
	Solar irradiance incident on a window	WinSol	$I_{g,w} > 100\ W/m^2$
Control strategy of opening pivoted windows	Percentage of the opened window area	Close Open	0% 100%
Control strategy of opening double-leaf windows	Percentage of the opened window area	Close Half Open	0% 50% 100%

Table 1.2 collects all the available options for controlling solar shading devices, and window opening during summer nights. Finally, to provide a good indoor air quality inside the building, a constant air change rate (ACR) of 0.6 h^{-1} is set in all simulations as a constraint. This value has been calculated according to EN 15251 [11].

Considering all the permutations of options available for each of the design variables, the problem space results constituted by 17 006 112 building variants, which cannot be simulated in a reasonable and feasible time span with a conventional approach. An optimization approach is therefore fundamental to reduce simulation times, converging faster toward the desired objective.

In order to run the optimization, the Particle Swarm Optimization (PSO) algorithm is selected, due to its robustness and efficiency to converge towards the global minimum, even if it requires more time than deterministic algorithms [29]. PSO is a heuristic population-based optimization algorithm and its development has been inspired by the social behaviour of animals. In the PSO semantic, the building variants simulated at each generation are called particles. They move in swarm through the problem space searching for the optimal solution. In the case of a minimization problem, the swarm of particles aims at such areas characterized by lower values of the objective function (or utility function).

The speed, with which each particle moves, depends on its best value compared with the overall best value found in the particle neighbourhood. The setting parameters used in the presented optimization run are:

- The type of algorithm is the PSO with inertia weight;
- The neighbourhood topology set as von Neumann;
- The neighbourhood size set to 5;
- The number of particles set to 20;
- The number of generations set to 30;
- The cognitive acceleration set to 2.8;

- The social acceleration set to 1.3;
- The initial inertia weight set to 1.2;
- The final inertia weight set to zero.

The total number of simulation runs for optimization is 600. The total runtime of optimization for the house has been in the order of 13 h on an iMac quad-core i5 at 2.5 GHz with 4 GB DDR3 RAM.

1.4.2 Dynamic Building Performance Simulation

The simulation engine adopted for computing the performance of the building is EnergyPlus [25] version 6.0.0.23. EnergyPlus is validated against the ANSI/ASHRAE 140 [30] and IEA SHC Task34/Annex43 BESTest method. Within the capability of EnergyPlus, the physical models and numerical algorithms for calculating heat exchanges have been selected making a trade-off between precision and computation time. The update frequency for calculating sun paths is set to 20 days, while the heat conduction through the opaque envelope is calculated via the conduction transfer function method with four time steps per hour. The natural convection heat exchange near external and internal surfaces is calculated via the adaptive convection algorithm [31] to better meet the local conditions of each surface of the model. The initialization period of simulation is set at 25 days, instead of the default value of seven days, to reduce the uncertainties connected to the thermal initialization of the numerical model. The voluntary ventilation and involuntary air infiltration are calculated with the *AirflowNetwork* module, instead of the simpler scheduled approach, to better estimate the contribution of natural ventilation and infiltration.

Adopting the optimization procedure described in detail in previous works [32–34], the building model is simulated in free-running mode during the whole year, and reference thermal comfort conditions are set according to the ASHRAE adaptive thermal comfort model [35] as expressed in ASHRAE 55 [36]. A critical review of comfort survey scales, comfort models and their application in building design is given in [37].

Summer and winter periods have been characterized using the method proposed in a dedicated publication by Carlucci [29]. For this specific case study, summer conventionally extends from June 1st to October 15th, and winter from November 15th to April 30th.

The numerical model includes also the Earth-to-Air Heat Exchanger (EAHE) designed in order to improve the thermal comfort performance of the building. The EAHE is constituted by 3 pipes spaced 1 m each and placed at a minimum depth of 3 m with a slope of 2.5%. It is spaced 4.7 m from the boundary surfaces of the lot and 3.5 m from the basement level of the building. The internal and external diameters of the pipes are respectively 142 mm and 160 mm. The fan is placed in the basement between the ground exchanger and the distribution system in the house. Its nominal airflow rate, at the maximum speed, is 350 m³/h. Further information about the sizing procedure for the EAHE are reported in Sect. 1.6. Because the model for EAHE integrated in EnergyPlus simulates only a single-tube configuration, a different modelling approach developed in a previous Ph.D. work [38] is adopted. It consists in coupling the EnergyPlus dynamic simulation of the building and a steady periodic simulation of the EAHE with the surrounding ground. The information used to simulate the EAHE are: (i) the outdoor dry-bulb air temperature taken from the weather file of the location used in the dynamic simulation of the building, (ii) the thermal properties of the ground, such as the diffusivity of the soil around the pipes and the depth of the aquifer, (iii) the layout of the EAHE (number of the pipes, position of the fan), (iv) the geometrical description of the pipes (length, diameter, distance from other pipes), (v) the distance from the building, (vi) the nominal air flow of the fan at the maximum speed, and (vii) the pressure drop of the EAHE.

The energy performance of EAHE is calculated assuming that:

- The convective flow inside the pipes is hydro-dynamically and thermally developed;
- The temperature profile in the pipe vicinity is not affected by the presence of the pipe. As a result, the pipe surface temperature is uniform in the axial direction;
- The soil surrounding the pipe is homogeneous and has a constant thermal conductivity;
- The pipes have a uniform cross sectional area in the axial direction.

The steady periodic regime allows calculating the hourly dry-bulb temperature and humidity ratio of outlet air from the EAHE for the whole year. They are subsequently set in EnergyPlus as temperature and humidity schedules for two coils (one for cooling and one for heating) and one humidifier. Therefore, in the model, outdoor air (characterized by the temperature and humidity of the weather file) is cooled to very low temperatures, then it is heated and humidified to achieve the temperature and humidity target conditions described by the two schedules calculated using the steady periodic regime model of the EAHE. A simple control rule is implemented for enhancing the summer performance: when outdoor air temperature is lower than the dry-bulb temperature of outlet air from the EAHE, then, the EAHE is bypassed, thus reducing the pressure losses and electric energy consumption for ventilative cooling.

1.4.3 Simulation Outcome and Discussion

Figure 1.5 depicts the outcome of the optimization run. Each point represents the thermal comfort performance of a given simulated building variant. The proposed optimization process has identified several building variants providing both winter and summer LPD lower than 10% (Fig. 1.5).

Fig. 1.5 Outcome of the
optimization run

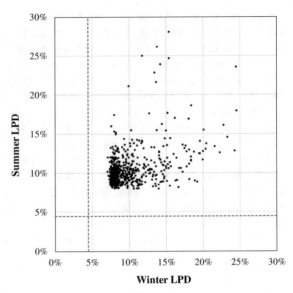

Simulated building variants

----Minimum LPD at theoretical comfort temperature

Moreover, the PSO algorithm identified an optimal building variant (x*),
according to the stated optimization problem, that achieves the lowest values of the
utility function $U(x^*)$. The values assumed as design variables of such optimal
building variant are:

- Very low values of steady-state transmittance of external walls and the roof,
  $U = 0.15$ W/(m^2 K); useful to limit heat exchange with outdoor in both the
  seasons;
- Relatively high value of steady-state transmittance of the floor, $U = 0.40$ W/(m^2
  K); it implies that the basement constitutes a heat sink during summer without
  compromising excessively winter performance;
- High value of time shift ($S > 12$ h) for the roof and the floor and a lower value
  of time shift (8 h $< S <$ 10 h) for external walls;
- Very low values of transmittance, $U_g = 0.59$ W/(m^2 K) and solar factor,
  $g = 0.36$, of glazing units on (almost) every orientation; which both reduce
  uncontrolled heat exchange through glazing; only glazing units of windows
  facing southeast (which is characterized by large glazing surfaces in this
  building) have a slightly higher solar factor, $g = 0.49$, which help enhancing
  solar gain during winter;
- The maximum value of the open window area during summer nights in order to
  provide maximum night free ventilative cooling;
- The control based on solar irradiance (WinSol), is the strategy selected to
  operate solar shading devices, but it would be useful to test more threshold
  alternatives.

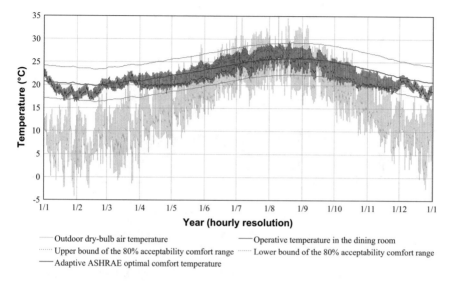

Fig. 1.6 Operative temperatures inside the living room in free-running mode compared with the 80% acceptability range of the ASHRAE adaptive model

The optimal building variant predicts indoor operative temperatures compatible with the 80% acceptability class of ASHRAE 55 in free-running mode; only few deviations occur outside the comfort zone (Fig. 1.6).

The energy use breakdown is: (i) 3.1 $kWh_{el}/(m^2$ a) for ventilation; (ii) 6.5 $kWh_{el}/(m^2$ a) for lighting; (iii) 15.3 $kWh_{el}/(m^2$ a) for electric equipment; (iv) 2.6 $kWh_{el}/(m^2$ a) for the production of domestic hot water. The annual required electricity is 4 087 kWh_{el}. Consumption due to electric equipment is relatively high since the house is also used as a small office 5 days per week and 8 h per day.

13 PV mono-crystalline panels (with a nominal efficiency of 18.4% and a peak power of 300 W per panel), with an overall DC to AC derate factor of 0.77) installed southwest facing on an area of 21.2 m^2, offer a nominal peak power of 3.9 kW_p and generate 4911 kWh_{el} per year (the slope of the roof is 22°).

Considering the balance over a year, the expected on-site electricity generation should be slightly higher than the whole electrical use, including lighting and electric appliances (Fig. 1.7). However, there could be a mismatch between generation and self-consumption on monthly, daily and hourly basis.

A second scenario has been considered, including the installation of a mechanical heating and cooling system (e.g., a reversible heat pump) into the concept of the building. Under this hypothesis, requirements about thermal comfort in indoor spaces have been set referring to the Fanger comfort model as implemented in the International standards ISO 7730 [39]. Since LPD provides a coordinated ranking when used both with the adaptive thermal comfort models and with the Fanger model [28, 40], the change does not heavily affect the ranking of the

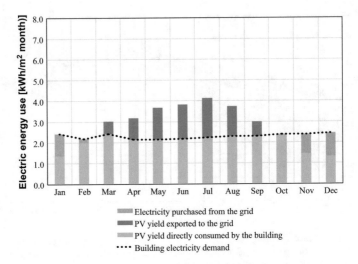

Fig. 1.7 Simulated electric energy balance of the house in free-running mode including PV yield

simulated building variants, i.e. the best building variant identified when requirements are defined according to the ASHRAE adaptive comfort model is the same to
that identified when using the Fanger comfort model.

In order to calculate seasonal optimal comfort temperatures according to the
Fanger comfort model, it is assumed a metabolic activity of occupants of 1.2 met, a
fixed summer clothing resistance of 0.5 clo, a fixed winter clothing resistance of 1.0
clo, an air velocity of 0.1 m/s, a target relative humidity of 50% and an external
work set at zero met. According to these assumptions the optimal summer operative
temperature is 26.5 °C and the optimal winter operative temperature is 21.2 °C.
The boundary temperatures of the comfort range are calculated in compliance with
the Category II of EN 15251 suitable for new buildings.

In Fig. 1.8, the annual fluctuation of indoor operative temperature inside the
dining room is drawn, contrasted to the outdoor (dry-bulb) air temperature and the
optimal comfort temperature and the Category II comfort ranges.

According to this scenario, the building is all-electric and the breakdown of
energy uses is the same either expressed in terms of delivered energy or primary
energy (Fig. 1.9).

Annual energy use for space heating amounts to 7.3 kWh/(m^2 a) and for space
cooling (sensible plus latent) is 9.5 kWh/(m^2 a). The overall electricity demand,
which includes space heating and cooling, dehumidification, production of
domestic hot water, ventilation, lighting, plug loads, is 7253 kWh per year, i.e.
48.8 kWh/(m^2 a).

Then, since the slope of the roof is 22° and assuming to install southwest facing
mono-crystalline cells with a nominal efficiency of 18.4% and a peak power of
300 W per panel, and an overall DC to AC derate factor of 0.77, 20 PV panels are
sufficient to balance (over one year) the whole electricity demand of the building.

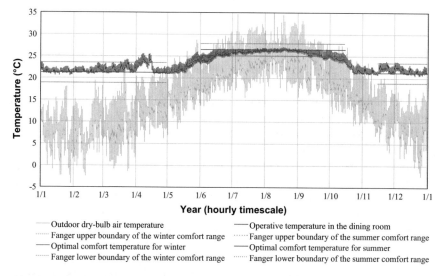

Outdoor dry-bulb air temperature   Operative temperature in the dining room
Fanger upper boundary of the winter comfort range   Fanger upper boundary of the summer comfort range
Optimal comfort temperature for winter   Optimal comfort temperature for summer
Fanger lower boundary of the winter comfort range   Fanger lower boundary of the summer comfort range

Fig. 1.8 Operative temperatures inside the living room in conditioned mode contrasted with the Category II range of the Fanger model and outdoor air temperature

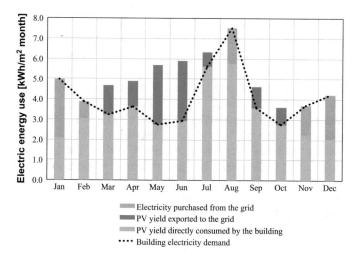

Electricity purchased from the grid
PV yield exported to the grid
PV yield directly consumed by the building
•••• Building electricity demand

Fig. 1.9 Simulated electric energy balance of the house including PV yield

The PV field is characterized by a nominal peak power of 6.0 kW$_p$ and a covered area of 32.6 m^2. The expected annual PV yield is 7580 kWh per year, hence, the building is expected to produce more electricity than it requires (Fig. 1.9).

Results show a good potential of the proposed passive concept (high insulated building coupled to an EAHE and internal thermal mass exploitation) in the selected climate, since the building is expected to produce, on-site, slightly more energy than all energy requirements and to guarantee high thermal comfort conditions.

1.5   Experimental Set-up

Building energy performance and thermal comfort conditions are monitored by a detailed monitoring system. The zero-energy living lab is intended as a full-scale-test building, operating under real conditions in the Mediterranean climate. The building automatic control system will allow to evaluate different strategies for solar protection and mechanical ventilation operation, including the EAHE, and considering the real dynamic effects of heat storage in building components and in the ground.

Monitoring will include: indoor air and operative temperatures, thermal energy demand for heating, cooling, and domestic hot water, electrical energy use for lighting and electrical equipment, the energy production by solar thermal and photovoltaic systems. The total primary energy demand will furthermore be calculated. Temperatures, water flow rates and thermal energy are measured in supply and return pipes via ultrasonic heat energy meters mounted in each hydronic loop of the system, with proper setting path upstream and downstream from the meters, to reach their best accuracy. Also electric energy demand for the reversible air to water heat pump and auxiliary systems is monitored via electric energy meters. The state and opening position of automatic mixing valves and of ventilation dampers is logged too.

A detailed monitoring layout [41] has been designed for the EAHE (Sect. 1.6) with several sensors for temperature (PT-100, class A accuracy, 4 wires connection), water content in the ground, and accurate air velocity measurements. Similarly, the monitoring of the air-to-air heat recovery system is performed with the installation of temperature, relative humidity and air velocity sensors, together with electrical meters for the fans.

Electric energy demand is separately monitored for lighting, domestic electrical equipment (oven, induction hob, refrigerator, dishwasher, washing and dryer machine, coffee machine), plug loads, actuators, and all the auxiliaries such as pumps and fans. The measurements may give detailed information also on internal heat gains related to lighting and electrical uses, which are estimated to be relevant in the energy balance of a passive building like this.

Important information on solar thermal gains and daylighting conditions are provided by monitoring of vertical position and slats angles of external blinds devices, which are automatically controlled for each windows.

Windows opening are logged to monitor where, when and how natural ventilation is adopted for free cooling. This kind of measures will allow to elaborate a long term monitoring of the occupants behaviour and of the BACS.

Thermal and visual comfort and indoor air quality are monitored by globe-thermometers, air temperature, relative humidity, CO_2 and lux sensors in the main rooms. Calculations of long-term comfort indexes will be performed and different methods will be tested and analysed. The use of globe-thermometers will allow to control thermal comfort on the basis of the actual operative temperature in contrast with the frequent approximation of assuming operative coincident with just air temperature.

In order to correlate building performance with external conditions, a weather station is installed on the roof of the building, measuring outdoor air temperature, wind velocity, external illuminance level, solar radiation and sun position. Two sensors close to the supply ducts of the EAHE further monitor external temperatures, and three sensors are placed at different heights in the patio, with the aim to study the cooling effect due to cool air falling from the sloped roof during clear summer nights. Outdoor relative humidity is also measured in the patio and close to supply ducts of the EAHE.

Different control strategies for natural and mechanical ventilation will be tested and monitored, particularly focusing on thermal comfort effect due to free cooling, when possible.

1.6   Earth to Air Heat Exchanger

EAHE represent a relatively simple technology that can be easily coupled with traditional HVAC systems as well as with hybrid systems in order to exploit the thermal capacity of the ground. The air is pre-heated in the heating season and pre-cooled in the cooling season through the EAHE, thus smoothing the peak loads and reducing the primary energy consumption. The experimental set-up of the EAHE implemented in the living lab is presented in a dedicated section because it is a passive technology that shows a substantial potential of application in the Mediterranean climate. One of the aims of the living lab is to characterize this technology, providing experimental data to be used for validation of existing analytical models and numerical tools, and possible improvements.

Due to the elevate thermal inertia of the ground, temperature fluctuations are much smaller below the ground than at surface level. At a sufficient depth, soil temperature is lower than the outdoor air temperature in the cooling season and higher in the heating season [42]. Thus, passing through the EAHE, the outdoor air can be effectively pre-heated or pre-cooled before entering the ventilation system [43, 44].

1.6.1   Location of the Earth-to-Air Heat Exchanger and Identification of the Boundary Conditions

Some of the features of the EAHE are bound by the geometric limits of the lot and the soil type. Figure 1.10 shows the layout of the house and the relative position of the underground pipes of the EAHE. The L-shape was chosen to respond to various requirements: (i) to reduce the thermal influence of the building and of the lot

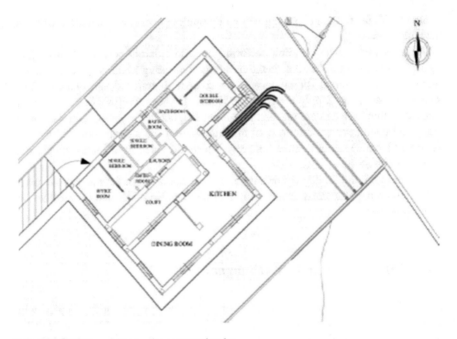

Fig. 1.10 EAHE position on the construction lot

boundary walls on the operation of the EAHE, (ii) to connect the pipes to the conveyor box of the ventilation system and (iii) to ensure easy periodic maintenance of the pipes.

Two of the boundary walls of the lot (on the northeast for the whole length and on the southeast for about 5 m starting from east), and the wall adjacent to the stair leading to the basement level are exposed to outdoor air. These boundary surfaces are influenced by outdoor air temperature and might negatively interfere with the operation of the EAHE. They were therefore thermally insulated by means of insulating panels installed on these walls before filling back the excavation with soil.

1.6.2   Design of the EAHE

The design parameters of an EAHE are the type of the backfill soil material (Tables 1.3 and 1.4), the characteristics of the pipes (depth of the bed, length, spacing, number and section diameter) and the nominal airflow of the fan.

Table 1.3 Physical properties of scoriaceous lava originally in place

Physical quantity	Value	Unit of measure
Thermal conductivity	1.28–2.79	W/(m K)
Specific heat	1150	J/(kg K)
Density	2600	kg/m^3
Thermal diffusivity	4.3×10^{-7}–9.3×10^{-7}	m^2/s
Daily periodic penetration depth	0.33–0.48	m
Annual periodic penetration depth	2.07–3.06	m

Fig. 1.11 The fine sand with clay used around the pipes (*first on the left*), the scoriaceous lava in fine-medium particles (*second on the left*), large stones (*third*) and topsoil (*fourth*)

Table 1.4 Physical properties of soils

Soil type	Density, ρ [kg/m^3]	Conductivity, λ [W/(m K)]	Specific heat, c [J/(kg K)]	Diffusivity, $\alpha = \lambda/\rho c$ [m^2/s]
FSC (dry)	1900	1.50	920	~8.56E-7
FSC (wet)	1900	2.50	1200	~1.10E-6
Scoriaceous lava	2600	1.28–2.79	1150	4.3E-7–9.3E-7

1.6.2.1   Selection of the Backfill Soil Material

All pipes were laid out and covered with a layer of low particle size (e.g., fine sand) that serves to compact the soil and increase the contact surface with the pipes [45]. Three types of soil have been used to cover the pipes: a mixture of fine sand and clay (FSC), scoriaceous lava (fine-medium particles and large stones) and topsoil (Fig. 1.11 and Table 1.4).

The FSC has been used for the entire section of the excavation, from 0.3 to 0.5 m below the pipes. It has been selected since it retains water and is not excessively draining and because it is characterized by fine particle size. The thermal conductivity of soil increases with the water content, therefore, a higher water content in the soil will result in a better heat exchange with the pipes.

The fine particles of the FSC allow to be compacted and to obtain a high-density soil, which on the one side improves the heat exchange with the surface of the pipes, and on the other ensures a better mechanical protection by uniformly distributing the soil pressure on the pipe surface. Outside the excavation site, the ground is constituted by scoriaceous lava that was the original soil of the whole lot. It consists of fine-to-medium particles and of large stones. The topsoil (typically used for gardens) is laid above the FSC layer for the entire surface of the garden around the building, in order to restore the original level of the garden and to allow gardening and planting.

1.6.2.2 Sizing of the EAHE

In order to both maximize the energy performance and guarantee a reliable comparison between measured and predicted data, the layout of the EAHE required a precise design. The underground pipes were installed at about 3 m depth, with a constant downward slope of 2.5%. They were connected to a conveyor box placed inside the utility room at the basement level. The slope is intended to allow for the drainage of water condensate.

An optimal functioning of the system requires that external (building and garden boundaries) as well as internal (mutual interactions of the pipes) heat perturbations are minimized. In order to guarantee both heating and cooling effective operation, the pipe spacing has been chosen to be more than twice the daily periodic penetration depth of the soil (Table 1.3), while the extension of the ground around the EAHE has been chosen to be larger than at least the annual periodic penetration depth. The periodic penetration depth, δ, is the "depth at which the amplitude of the temperature variations are reduced by the factor "e" (the Euler number equal to about 2.718) in a homogeneous material of infinite thickness subjected to sinusoidal temperature variations on its surface" [24]. Assuming a periodic regime and the homogeneity of the FSC soil, the daily and annual periodic penetration depths (shown in Fig. 1.12) are considered in order to optimize the EAHE layout (Figs. 1.13 and 1.14).

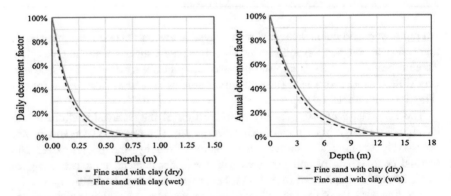

Fig. 1.12 Decrement of the heat wave amplitude as a function of depth of the FSC layer for daily (*left image*) and annual oscillations (*right image*)

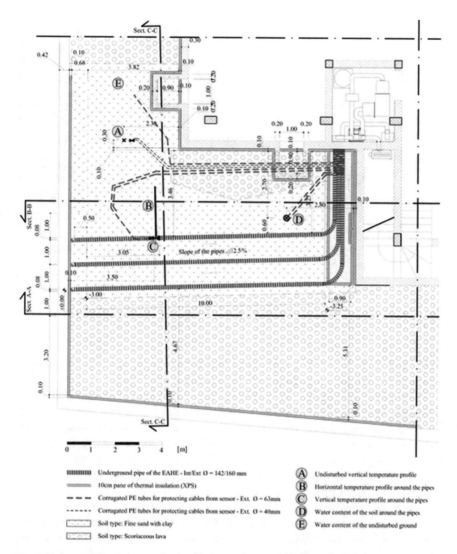

Fig. 1.13 Plan view of the EAHE with the indication of (i) the actual position of the EAHE, (ii) the measuring points and (iii) the types of the backfill soil

The daily periodic penetration depth of the FSC soil is 0.17 m if it is wet (and 0.15 m if it is dry), which means that a pipe spacing of 1 m reduces the amplitude of the daily oscillation by 95%. The same spacing of 1 m was suggested also by Zimmermann and Remund [46]. The extension of the ground around the pipes should be larger than at least the annual periodic penetration depth, which is about 3.3 m for the wet FSC soil. A similar choice for pipe depth was made by Ascione,

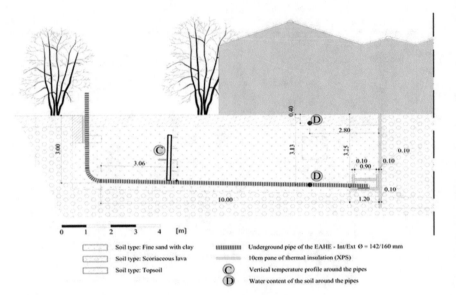

Fig. 1.14 Section A-A (see Fig. 1.13) of the EAHE with the indication of (i) the actual position of the EAHE, (ii) the measuring points and (iii) the types of the backfill soil

Bellia and Minichiello [45], who found that the best compromise between excavation costs and system performance was around 3 m.

Hollmuller and Lachal [47] discuss the daily amplitude-dampening (the day/night meteorological extremes around the daily average) and the yearly amplitude-dampening (the daily average approaching the yearly meteorological average) achievable using the thermal inertia of the ground. They analyse and contrast two geometries, (a) with pipes buried deeply and wide apart and (b) with pipes buried close to the surface and to each other, with approximately 20 cm soil around each. On the assumption that the boundary conditions at the soil surface are adiabatic and referring to the climate of Geneva, they conclude on the possibility of summer precooling with configuration (b), which has obviously a lower installation cost. Apart from the needed correction for the actual non-adiabatic boundary conditions, the summer weather in Mascalucia is such that aiming at just the dampening of the daily oscillations would not be sufficient to achieve, at the output of the EAHE, air temperatures within the comfort condition range. It would be even more difficult to effectively remove the cooling load due to solar radiation, occupants etc. Hence, in this climate and for summer cooling, there is the need for a location of the tubes relatively deep in the soil, in order to take profit of the yearly dampening. The availability of pre-heated air in winter is a bonus connected with this solution, although this result in isolation would not be economically competitive with a high efficiency heat recovery unit.

Compatibly with the lot layout, the pipes have thus been laid as distant as possible from external interfering surfaces (Figs. 1.13 and 1.14). The surfaces and the boundary walls were insulated with 5 cm of high-density polystyrene in order to minimize un-controlled heat exchanges.

1.6.2.3 Selection of the Pipe Diameter and the Nominal Airflow of the Fan

The mechanical ventilation system of the house is equipped with a highly efficient heat recovery unit ($\eta th = 0.95$), a variable speed fan and a by-pass unit. Potentially available variants are obtained combining two sizes of the fan (nominal air flow 350 m³/h and nominal electric power 105 W; or 550 m³/h and 350 W) and two sizes of the pipes (nominal external-internal diameters: $D_e = 110$ mm— $D_i = 93$ mm or $D_e = 160$ mm—$D_i = 142$ mm).

Since the net volume of the building is 443 m³, the two sizes of the fan nominally provide air change rates respectively of 0.8 and 1.2 h⁻¹. Four variants of the ventilation system coupled to the EAHE can be obtained varying the size of the fan and the section of the pipes. The energy performance of these variants were investigated with the software GAEA version 1.4.05 [48], using as inputs the geometric features of the EAHE, the features of the fan and the thermo-physical properties of the FSC soil put in place (Table 1.4). For all solutions the pipe length was maintained constant. Each pipe consists of a straight Sect. 10 m long and a bent section of 4 to 6 m that ends in a collector box. According to the manufacturer, the average efficiency of the fans (for both sizes) is 0.7. The total pressure drop was calculated for the four variants considering two 90° bends for each pipe (Table 1.5).

The electric energy absorbed by the fan and the cooling effect provided by the EAHE (in terms of heat subtracted from the warm, outdoor air) were simulated with the software GAEA, in the period from 1st May to 30th October. Hence, the Energy Efficiency Ratio (EER) was derived as the ratio between the thermal energy removed from the outdoor air by the EAHE during the calculation period and the electric energy used to operate the fan for the four variants (Table 1.6).

By-pass strategies were not included in such calculations; the comparison among several control strategies of the by-pass can be found in a former article [49].

Table 1.5 Pressure drop for the four available variants

	Variant 1	Variant 2	Variant 3	Variant 4
Nominal air flow of the fan [m³/h]	350	350	550	550
Diameter of pipes: Ext|Int [mm]	110|93	160|142	110|93	160|142
Dynamic pressure drop [Pa]	42.2	7.8	94.9	17.5
Distributed pressure drop [Pa]	162.1	21.7	331.1	44.1
Local pressure drop [Pa]	29.9	5.5	67.4	12.4
EAHE-side pressure drop [Pa]	234.2	35	493.4	74
House-side pressure drop [Pa]	28	28	28	28
Total pressure drop [Pa]	262.2	63	521.4	102

Table 1.6 Energy performance of the four variants of the mechanical ventilation system coupled to the EAHE

Physical quantity [Unit of measure]	Variant 1	Variant 2	Variant 3	Variant 4
Nominal air flow of the fan [m³/h]	350	350	550	550
Diameter of pipes: Ext|Int [mm]	110|93	160|142	110|93	160|142
Energy removed by the EAHE [kWh]	944	890	957	1035
Electric energy absorbed by the fan [kWh]	95	31	281	69
Energy efficiency ratio (EER) [-]	9.9	29.2	3.4	14.9

According to the assumptions mentioned, Variant 4 offers the largest potential to meet the energy need for space cooling, but it requires more than twice the electricity required by Variant 2, which is able to remove energy in a slightly lower quantity (14% less). Variant 2 (350 m³/h, $D_e = 160$ mm) has hence the highest EER and thus is the one finally chosen.

1.6.3 Design of the Monitoring System of the Earth-to-Air Heat Exchanger

In order to assess the effective energy performance of the EAHE based on long-term measurements, a monitoring system was installed. Both temperature (by means of PT- 100, class A sensors) and soil water content are monitored at different points (Figs. 1.13 and 1.14):

- Point A: The vertical temperature profile of the undisturbed ground is measured at the depth of 1.0, 1.5, 2.0, 2.5, 3.0 m;
- Point B: The horizontal temperature profile of the ground around the pipes is measured at a depth of 2.5 m and at distance from the external surface of the pipe of 0.05, 0.10, 0.15, 0.20, 0.40, 0.80, 1.20, 1.95 m;
- Point C: The vertical temperature profile of the ground around the pipes is measured at the depth of 1.0, 1.5, 2.0, 2.5 m. The surface temperature of the pipe is measured by dedicated devices, using a highly conductive adhesive to reduce the contact resistance;
- Point D: The volumetric soil water content is measured in the area around the pipes by means of two sensors that use the 'Frequency Domain Reflection' method, placed at the depth of 0.5 and 3.1 m. The latter has been located at the same depth of the pipes;
- Point E: The volumetric soil water content is measured in the undisturbed ground by means of one sensor that uses the 'Frequency Domain Reflection' method, placed at the depth of 0.5 m.

1.6.4 Installation of the Earth-to-Air Heat Exchanger

The EAHE was installed in 2011, after the building structure was completed. The installation phases were: (i) excavation of the ground, (ii) insulation of the perturbing boundary surfaces, (iii) creation of the pipe bed, (iv) pipe laying, (v) installation of the measuring instruments, (vi) covering of the pipes and finally (vii) testing of the sensors after installation.

In order not to interfere with the thermal field around the soil temperature sensors, an ad hoc supporting structure was created. A H-shape fiberglass beam ($\lambda = 0.2$ W/(m K); $\rho = 1\ 900$ kg/m^3; c = 1.0 kJ/(kg K)) was selected for its thermal and structural properties. Sensors were installed on it in a laboratory, according to the indications reported by manufacturers (Fig. 1.15).

The 4-wire cables connected to the sensors were protected by corrugated tubes and laid up to the utility room, where an interface with a data acquisition system was placed. Before backfilling the excavation, a test of the sensors was performed. The electrical resistance between red and white clamps was measured on site for each sensor. This test is intended to check whether sensors and connecting cables have been damaged during transport or installation: a damaged sensor usually shows an out-of-range value of electrical resistance. The test run was successful for all sensors. The following phases were: (i) covering (manually) the pipes and the blue corrugated tubes with shrunken sand-and-clay soil; (ii) controlling the verticality of the H-shape support; (iii) backfilling up to a depth of 0.5 m with FSC soil; (iv) restoring the garden surface with the original topsoil. Some soil samples were collected in order to measure their equivalent conductivity in laboratory.

Fig. 1.15 H-shape fiberglass beam with sensors (*left* and *centre*), protection of the cables by means of *blue* corrugated tubes (*centre*), and installation of *horizontal sensors* (*right*)

1.7   System Start-up and Early Outcomes

The building construction was concluded in 2013, nevertheless during 2014 additional work was necessary to complete the detailed monitoring system and the installation of the building automation and control system. Meanwhile, a family of three persons occupied the building, and building systems and operable windows were relying on occupant control.

The commissioning phase included the EAHE, the heat pump and all the components of the heating/cooling and ventilation systems. The position of the monitoring equipment was optimized in order to have the best response according to operational data gathered during the commissioning phase. Also the photovoltaic and thermal solar panel systems were commissioned and their operation was carefully controlled.

The layout of the indoor climate monitoring system was defined according to the real use of the building, avoiding disturbance with occupants, which would otherwise determine bias in the outcomes.

The basic algorithms for the control of the heating/cooling and ventilation systems were included in the BACS after the commissioning phase. More detailed control logics including lighting, solar shading and ventilation are under development and will shortly be tested under real operational conditions.

Some early outcomes are presented in Figs. 1.16 and 1.17, showing temperature and relative humidity values from the 23rd to the 30th June 2014, in three reference

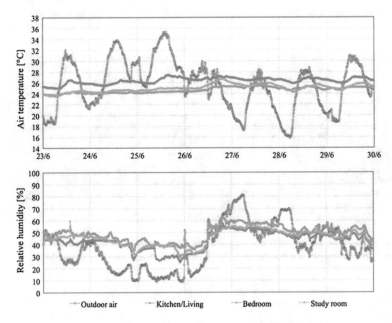

Fig. 1.16 Temperature and relative humidity values recorded in the kitchen, bedroom and study of the building from 23rd to 30th June 2014

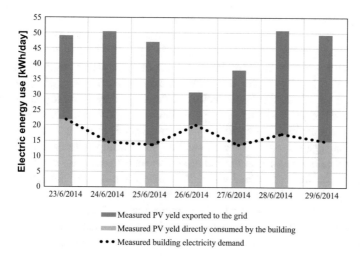

Fig. 1.17 Daily total electric energy use of the building vs. total electric energy generated by the PV system from 23rd to 30th June 2014

rooms (Fig. 1.16) and the daily electricity use and generation during the same days (Fig. 1.17). The data refers to a period when the building was operated under free running condition, i.e. no mechanical cooling was provided. Solar shading and lighting were under the occupant control and no automatic control was provided.

The building shows good environmental conditions also in the kitchen where the highest solar and internal heat gains are experienced. The electrical energy generated by the PV system substantially overcomes the energy use.

1.8   Conclusions

In order to reduce the gap existing between operational and design energy performance of buildings, analyses based on real monitoring data are required. Among the available facilities, living labs, i.e. full scale buildings acting as experimental laboratories exposed to real weather, may provide useful information on the global operational performance of a building, allowing to test the effectiveness of different technologies for the reduction of energy consumption or the enhancement of indoor environmental quality.

An advanced living lab has been designed and constructed in Mascalucia (Italy) in order to test the possibilities and limitation of passive and zero-energy homes in the Mediterranean climate. The design process adopted advanced optimization techniques, coupled with energy simulation in order to provide design solutions, which could minimize energy consumption while achieving high thermal comfort levels. Operational data will provide information to test simulation results, and eventually to improve the virtual model and the simulation techniques. The earth to air heat exchanger will be monitored with special detail and will hence deliver data of high quality to be used for further validation and improvement of EAHE numerical models (e.g., those implemented in TRNSYS and EnergyPlus).

The living lab was built following the energy simulation results, which show evidence of a high potential for a zero-energy balance, when adequate passive strategies are adopted. The house includes a detailed energy monitoring system, that will allow to define an accurate energy breakdown for the building. Also indoor and outdoor environmental parameters will be monitored, in order to contrast energy performance against environmental quality. A building automation and control system will allow to operate the building also under automatic control, providing information on the operational effect of human control and automatic control of building systems, including solar screens and ventilation.

The early outcomes, following the commissioning phase, show that the indoor thermal conditions under free running conditions (no mechanical cooling) match fairly well the simulation results. The building appears able to guarantee adequate comfort conditions adopting purely passive strategies. The early monitoring data of daily electrical demand and production correspond to what forecasted by energy simulation.

Longer measurements are nevertheless required to ascertain the performance of the building in terms of the indices identified in reference works [9] (energy needs, net yearly primary energy balance, load match and long-term thermal comfort indices), and to determine what effects occupant and automatic controls have on those indices. After the complete implementation of all the control algorithms into the BACS, a long and detailed monitoring campaign will start, that will provide extensive data on the operational condition of the building.

Acknowledgements The authors would like to thank all participants of IEA ECBCS Annex 58 entitled "Reliable Building Energy Performance Characterisation Based on Full Scale Dynamic Measurements" for the useful discussions, Ing. Carmelo Sapienza of Sapienza&Partners engineering firm, ROCKWOOL Italia and SIEMENS Building Technologies for their technical support. This research was partly conducted in the frame of PassREg project entitled "Passive House Regions with Renewable Energies" and supported by the Intelligent Energy Europe programme.

References

1. European Parliament and Council.: Energy Performance of Buildings Directive 2002/91/EC. Official Journal European Union, Luxemburg (2002)
2. European Parliament and Council.: Energy Performance of Buildings: Directive 2010/31/EU, vol. L. 153/13. Official Journal of the European Union, Luxembourg (2010)
3. Altwies, J.E., Nemet, G.F.: Innovation in the U.S. building sector: An assessment of patent citations in building energy control technology. Energy Policy **52**, 819–831 (2013)
4. ZCH.: Closing the gap between design & as-built performance—End of term report. Zero Carbon Hub (2014)
5. Ostime, N.: RIBA Job Book. RIBA Publishing, London (UK) (2013)

6. Sinclair, D.: Guide to Using the RIBA Plan of Work 2013. RIBA Publishing, London (UK) (2013)
7. Torcellini, P., Pless, S., Deru, M., Crawley, D.: Zero Energy Buildings: A Critical Look at the Definition. ACEEE Summer Study, Pacific Grove, California, USA (2006)
8. Marszal, A.J., Heiselberg, P., Bourrelle, J.S., Musall, E., Voss, K., Sartori, I., Napolitano, A.: Zero energy building—a review of definitions and calculation methodologies. Energ Buildings **43**, 971–979 (2011)
9. Hermelink, A., Schimschar, S., Boermans, T., Pagliano, L., Zangheri, P., Armani, R., Voss, K., Musall, E.: Towards nearly zero-energy buildings: Definition of common principles under the EPBD Ecofys. University of Wuppertal, Politecnico di Milano - eERG (2013)
10. CEN: Energy performance of buildings—Overall energy-use and definition of energy ratings. vol. EN 15603. European Committee for Standardization, Bruxelles, Belgium (2008)
11. CEN.: Indoor Environmental Input Parameters for Design and Assessment of Energy Performance of Buildings Addressing Indoor Air Quality, Thermal Environment, Lighting and Acoustics. vol. EN 15251. European Committee for Standardization, Brussels, Belgium (2007)
12. Carlucci, S., Pagliano, L.: A review of indices for the long-term evaluation of the general thermal comfort conditions in buildings. Energ Buildings **53**, 194–205 (2012)
13. Pagliano, L., Carlucci, S., Toppi, T., Zangheri, P.: Passivhaus per il sud dell'Europa - Linee guida per la progettazione. Rockwool Italia, Milano (2009)
14. end-use Efficiency Research Group. http://www.passive-on.org/en/
15. Causone, F., Carlucci, S., Pagliano, L., Pietrobon, M.: A zero energy concept building for the Mediterranean climate. Energy Procedia **62**, 280–288 (2014)
16. Peel, M.C., Finlayson, B.L., McMahon, T.A.: Updated world map of the Köppen—Geiger climate classification. Hydrol. Earth Syst. Sci. **11**, 1633–1644 (2007)
17. McKnight, T.L., Hess, D.: Climate Zones and Type. Physical Geography: A Landscape Appreciation. Prentice Hall, Upper Saddle River, NJ, USA (2000)
18. Köppen, W.P., Geiger, R.: Handbuch der klimatologie. Gebrüder Borntraeger, Berlin (1930)
19. https://en.wikipedia.org/wiki/Mediterranean_climate
20. Serghides, D.K., Georgakis, C.G.: The building envelope of Mediterranean houses: Optimization of mass and insulation. J. Build. Phys. **36**, 83–98 (2012)
21. de Saulles, T.: Thermal Mass Explained. MPA - The Concrete Centre, Camberley (2009)
22. Jaber, S., Ajib, S.: Optimum technical and energy efficiency design of residential building in Mediterranean region. Energ. Build. **43**, 1829–1834 (2011)
23. Santamouris, M., Kolokotsa, D.: Passive cooling dissipation techniques for buildings and other structures: The state of the art. Energ. Build. **57**, 74–94 (2013)
24. CEN.: Thermal performance of building components—Dynamic thermal characteristics—Calculation methods. vol. EN ISO 13786. European Committee for Standardization, Brussels, Belgium (2007)
25. Crawley, D.B., Lawrie, L.K., Winkelmann, F.C., Buhl, W.F., Huang, Y.J., Pedersen, C.O., Strand, R.K., Liesen, R.J., Fisher, D.E., Witte, M.J., Glazer, J.: EnergyPlus: Creating a new-generation building energy simulation program. Energ. Build. **33**, 319–331 (2001)
26. Wetter, M.: GenOpt—A Generic Optimization Program. Seventh International IBPSA Conference, pp. 601–608, Rio de Janeiro (2001)
27. Athienitis, A., O'Brien, W., Cellura, M., Ayoub, J., Pagliano, L., Hasan, A., Carlucci, S., Bourdoukan, P., Attia, S., Delisle, V., Candanedo, J.A., Hamdy, M., Bucking, S., Chen, Y., Guarino, F., Lenoir, A.: Modelling, Design, and Optimization of Net-Zero Energy Buildings. Ernst & Sohn, Berlin, Germany (2015)
28. Carlucci, S.: Thermal Comfort Assessment of Buildings. Springer, London (2013)
29. Hopfe, C.J.: Uncertainty and sensitivity analysis in building performance simulation for decision support and design optimization. Department of Architecture, Building and Planning, vol. Ph.D., pp. 215. Eindhoven University of Technology, Eindhoven, The Netherlands (2009)

30. ANSI/ASHRAE 140.: Standard Method of Test for the Evaluation of Building Energy Analysis Computer Programs. vol. ANSI/ASHRAE 140, pp. 272. American Society of Heating, Refrigerating and Air-Conditioning Engineers, Atlanta (GA), USA (2011)
31. US-DoE.: InputOutput Reference: The Encyclopedic Reference to EnergyPlus Input and Output. U.S. Department of Energy (2010)
32. Carlucci, S.: An automated optimization process to support the design of comfortable net zero energy buildings. Energy, vol. Ph.D. Italy, Milan (2012)
33. Carlucci, S., Pagliano, L.: An optimization procedure based on thermal discomfort minimization to support the design of comfortable net zero energy buildings. In: 13th Conference of the International Building Performance Simulation Association, BS 2013, pp. 3690–3697 (2013)
34. Carlucci, S., Pagliano, L., Zangheri, P.: Optimization by discomfort minimization for designing a comfortable net zero energy building in the mediterranean climate. Adv. Mater. Res. **689**, 44–48 (2013)
35. De Dear, R., Brager, G.S.: The adaptive model of thermal comfort and energy conservation in the built environment. Int. J. Biometeorol. **45**, 100–108 (2001)
36. ANSI/ASHRAE 55.: Thermal Environmental Conditions for Human Occupancy. vol. ANSI/ASHRAE 55. American Society of Heating, Refrigerating and Air- Conditioning Engineers, Atlanta, USA (2010)
37. Pagliano, L., Zangheri, P.: Comfort models and cooling of buildings in the Mediterranean zone. Adv. Build. Energy Res. **4**(1), (2010). http://doi.org/10.3763/aber.2009.0406
38. Zangheri, P.: Development of optimization procedures for integrated Building Envelope and Low Energy Cooling Systems using different comfort models and application to the Passivhaus concept. Energy Department, vol. Ph.D. Politecnico di Milano, Milan, Italy (2011)
39. ISO.: Analytical determination and interpretation of thermal comfort using calculation of the PMV and PPD indices and local thermal comfort criteria. Ergonomics of the thermal environment, vol. ISO 7730, pp. 52. International Organization for Standards, Geneva (2005)
40. Carlucci, S., Pagliano, L., Sangalli, A.: Statistical analysis of ranking capability of long-term thermal discomfort indices and their adoption in optimization processes to support building design. Build. Environ. **75**, 114–131 (2014)
41. Carlucci, S., Cattarin, G., Pagliano, L., Pietrobon, M.: Optimization of the installation of an Earth-to-Air Heat Exchanger and detailed design of a dedicated experimental set-up. Appl. Mech. Mater. **501–504**, 2158–2161 (2014)
42. Peretti, C., Zarrella, A., De Carli, M., Zecchin, R.: The design and environmental evaluation of earth-to-air heat exchangers (EAHE). A literature review. Renew. Sustain. Energy Rev. **28**, 107–116 (2013)
43. Pagliano, L., Zangheri, P., Carlucci, S.: A Way To Net Zero Energy Buildings For Italy: How The Earth-To-Air Heat Exchanger Could Contribute To Reach The Target In Warm Climates. EuroSun 2010—International Conference on Solar Heating, Cooling and Buildings, Graz (2010)
44. Carlucci, S., Zangheri, P., Pagliano, L.: Achieving the net zero energy target in Northern Italy: Lessons learned from an existing Passivhaus with Earth-to-Air heat exchanger. Adv. Mater. Res. **689**, 184–187 (2013)
45. Ascione, F., Bellia, L., Minichiello, F.: Earth-to-air heat exchangers for Italian climates. Renew. Energy **36**, 2177–2188 (2011)
46. Zimmermann, M., Remund, S.: Chapter F—ground coupled air systems. Low energy cooling—technology selection and early design guidance, Construction Research Communications Ltd. (2001)
47. Hollmuller, P., Lachal, B.: Air–soil heat exchangers for heating and cooling of buildings: Design guidelines, potentials and constraints, system integration and global energy balance. Appl. Energy **119**, 476–487 (2014)
48. Benkert, S.T., Heidt, F.D., Schöle, D.: Calculation tool for Earth heat exchangers GAEA. Building Simulation 1997—The fifth international IBPSA conference, Prague (CZ) (1997)

49. Zangheri, P., Pagliano, L.: Methodology for design and evaluation of zero energy buildings in mediterranean climate. Application to a Passivhaus with EAHE. 3rd Passive And Low ENergy Cooling for the built environment Conference (PALENC 2010) and 5th European Conference on Energy Performance & Indoor Climate in Buildings (EPIC 2010), pp. 1–10, Rhodes Island (GR) (2010)

Chapter 2
Assessment of the Green Roofs Thermal Dynamic Behavior for Increasing the Building Energy Efficiencies

Antonio Gagliano, Maurizio Detommaso and Francesco Nocera

Abstract The continuous city growth leads to the intensification of the Urban Heat Island effect (UHI) which involves several consequences for the in-habitant comfort. Countermeasures such as green roofs can mitigate the urban microclimate reducing the roof surface temperature and hence the surrounding air temperature. As most of the existing Italy building stock was built without or with poor thermal insulation, they need of interventions of energy retrofit. In particular, roofs are the component of the building envelope that are mainly engaged of both solar gains and thermal fluxes. Therefore, any interventions refurbishment should include the roof surface. This paper focuses on the analyses of the energy performance and the dynamic thermal behavior of a green roof implemented in an existing house-holiday situated in the south part of Sicily. Moreover, other two refurbishment scenarios that foresee the increase of the roof thermal insulation were analysed. Dynamic simulations have proven that the green roof reduce the energy needs, about 90% in the cooling period and 30% in the heating period; delays and attenuates the outdoor heat wave in comparison with traditional roofs, as well as diminishes the average daily temperature fluctuations. The reduction of the vegetated temperature area results about 10.0 °C lower than the one of a bare roof area.

Keywords Green roof · Energy building simulations · Thermal comfort · UHI

Nomenclature

$C_{e,g}$	Latent heat flux bulk transfer coefficient at ground layer
C_f	Bulk heat transfer coefficient
$C_{h,g}$	Sensible heat flux bulk transfer coefficient at ground layer
$C_{p,a}$	Specific heat (J kg^{-1} K^{-1})
DF	Decrement factor ($-$)

A. Gagliano (✉) · M. Detommaso · F. Nocera
Department of Electric, Electronics and Computer Engineering, University of Catania, via A. Doria 6, Catania 95125, Italy
e-mail: agagliano@dii.unict.it

© Springer International Publishing AG 2017
J. Littlewood et al. (eds.), *Smart Energy Control Systems for Sustainable Buildings*,
Smart Innovation, Systems and Technologies 67,
DOI 10.1007/978-3-319-52076-6_2

F_f Net heat flux to foliage layer (W m^{-2})

F_g Net heat flux to ground surface (W m^{-2})

h_c Convective heat transfer coefficient (W m^{-2} K)

h_r Linear radiative heat transfer coefficient (W m^{-2} K)

H_f Foliage sensible heat flux (W m^{-2})

H_g Ground sensible heat flux (W m^{-2})

$I_\downarrow$ Total incoming short wave radiation (W m^{-2})

$I_{ir}^\downarrow$ Total incoming long wave radiation (W m^{-2})

l_f Latent heat of vaporization at foliage temperature (J kg^{-1})

l_g Latent heat of vaporization at ground temperature (J kg^{-1})

L_f Foliage latent heat flux (W m^{-2})

L_g Ground latent heat flux (W m^{-2})

LAI   Leaf area index (m m^{-2})

PMV   Predicted mean vote

PPD   Predicted percentage of dissatisfied

q_{af} Mixing ratio for air

$q_{f,sat}$ Saturation mixing ratio at foliage temperature

$q_{g,sat}$ Saturation mixing ratio at ground temperature

$r_{..}$ Surface wetness factor

r Short wave reflectance (−)

r_f Short wave reflectance of the foliage (−)

r_g Short wave reflectance of the soil surface (−)

s Thickness (m)

S Total envelope surface area (m^2)

S_r Roof surface (m^2)

S_u Net floor area (m^2)

SM   Superficial mass (kg m^{-2})

t Transmissivity (−)

T_{af} Air temperature with in the canopy (°C)

T_{db} Dry bulb air temperature

T_f Leaf temperature (°C)

T_g Ground surface temperature (°C)

T_i Indoor temperature (°C)

T_{MRT} Mean radiant temperature (°C)

T_o Outdoor air temperature (°C)

T_{op} Opertive temperature (°C)

$T_{op,av}$ Average opertive temperature (°C)

T_{si} Inner surface temperature (°C)

T_{so} Outdoor surface temperature (°C)

TL   Time-lag (h)

U Thermal transmittance (Wm^{-2} K^{-1})

V Heated gross-volume (m^3)

W_{af} Wind speed with in the canopy (m/s)

z Depth (m)

Greek letters

α_f	Short wave absorptance of canopy (−)
α_g	Short wave absorptance of the ground surface (−)
ε_f	Emissivity of canopy (−)
ε_g	Emissivity of the ground surface (−)
ε_1	$E_f + \varepsilon_g - \varepsilon_f \varepsilon_g$
g	Radiant fraction
λ	Thermal conductivity (W m^{-2} K^{-1})
ρ_{af}	Density of air at foliage temperature (kg m^{-3})
ρ_{ag}	Density of air at ground surface temperature (kg m^{-3})
ρ_g	Density of growing medium (kg m^{-3})
ρ	Density (kg m^{-3})
	Stefan-Boltzmann constant (W m^{-2} K^{-4})
f	Fractional vegetation coverage (−)
τ	Time (s)

2.1 Introduction

The improvement of building energy efficiency is at the heart of the EU's transition to a resource-efficient economy and the Europe 2020 strategy for smart, sustainable and inclusive growth.

Energy efficiency measures contribute to the overall resilience to climate change as they play a safeguarding role against adverse/extreme weather events.

Thus, research on innovative energy technologies and passive solutions for improving the energy efficiency of buildings has become a strategic environmental and economic issue.

Various sustainable approaches and environmentally responsive efficient technologies have been proposed to design low-energy buildings [1–5].

In this context, the green roofs represents a suitable solution for improving the roof energy resilience in summer period reducing the cooling loads of both new and existing buildings [4, 6, 7] and for making buildings more sustainable.

The vegetation layer of a green roof realizes photo-synthesis processes, whereas the soil layer allows absorption of rainfall, often resulting in improvements of the water runoff quality [8–10]. Many researches on the green roofs have pointed out further benefits on the environment, such as improving thermal comfort and urban air quality [11], mitigation of the Urban Heat Island (UHI) by reducing "hot" surfaces facing the sky [6], [12–14] as well as noise transmission.

Green roofs are generally built to enhance the energy saving, since they act as a thermal buffer [15, 16] diminish the heat gain of about 70–90% in summer, and heat loss of about 10–30% in winter [17]. The energy consumptions of air conditioning (AC) during the summer are reduced by means of the combined effect of shading and evapotranspiration [18]. They also reduce and delay the outdoor surface temperature peak and keep the internal conditions within the comfort range [14, 19, 20].

Green roofs, as well as the cool roofs, could show a negative energy balance compared with traditional roofs in the heating period, particularly during spring and autumn when the thermal losses are higher than the traditional roof [19].

As regard mitigation strategies aiming to reduce UHI, the implementation of green and cool roofs have been proposed in literature [5, 6, 13].

However, the mechanisms to reduce UHI through green and cool roofs are different; a green roof acts by the evapotranspiration of soil and plants on rooftops (redirecting available energy to latent heat) [21], while a cool roof increases the reflection of incoming solar radiation by increasing the albedo of roof surfaces. The question of determining the impacts of these strategies of mitigation on the UHI phenomena at city scale remains an open question.

The purpose of this study is to evaluate the improvement of the energy performance of an existing poor insulated building, which has been refurbished by means of the implementation of an extensive averagely insulated green roof.

The research was conducted through dynamic simulations using the DesignBuilder software [22]. The energy performance of the building has been calculated under three different configuration roof configurations that foresee the implementation of a green roof, as well as the installation of two different thickness of thermal insulation (intermediate or well-insulated roofs). The dynamic thermal behavior, the thermal inertia parameters and the thermal comfort were investigated too. The results of the study show the reduction of energy needs, the improvement of the indoor thermal comfort and, the decreasing of the outdoor temperature surface when the green roof is applied.

2.2 Materials and Methods

Green roofs, also named "eco-roofs", "living roofs" or "roof gardens", are roofs with plants in their final layer [23, 24]. Two types of green roofs are generally identified: extensive (with soil thickness less than 10–15 cm) and intensive (with soil thickness more than 15–20 cm). As regard the intensive green roof, Karachaliou et al. presented a research on the thermal behavior and the energy performance of this typology of green roofs [25].

Because of their low additional loads, extensive green roofs are most suitable for building retrofitting, i.e. they do not require any additional strengthening [26].

The typical layers from the inside to the outside of an extensive green roofs are: load-bearing slabs (roof deck), thermal insulation, waterproof membrane, root barrier, drainage, filter layer, growing medium and vegetation layer (Fig. 2.1).

The vegetation layer can be constituted by mosses, sedum, graminaceous or other succulents plants, which are very common and suitable plants for using on an extensive green roof.

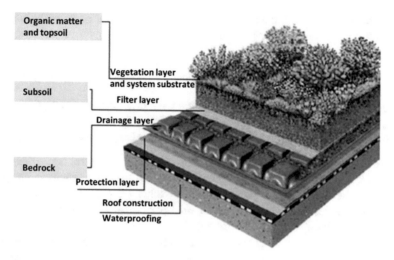

Fig. 2.1 Stratigraphy of extensive green roof [27]

All the above mentioned plants are small plants that grow across the ground rather that upwards offering good coverage and roof membrane protection. Moreover, such plants have the capabilities to store water in their leaves making them high drought resistance. Thus these extensive green roofs could survive also in Mediterranean climates. The systems must be able to retain the necessary quantities of water to support the plants, while draining off the excess. The plant type, the geographical region and the roof itself determine the required amount of water. Besides to build the most correct structure of green roof to support the plants, it is essential to protect the water-proofing from both mechanical damage as well as by the attack by plant roots.

The thermo-physical model of green roof identifies three main components: structural support, soil and canopy (leaf cover). The structural support includes all the layers between the inner plaster and the filter layer whose upper surface is the interface support-soil.

The soil is a complex component because it consists of a solid phase (mineral and organic material), a liquid phase (water) and a gaseous phase (air and water vapor). Data about soils are available in the technical standard EN ISO 13370 (CEN 2007) that provides thermal characteristics of common soils, as well as in the Farouki's database of soils [28].

Values of density, thermal conductivity, specific heat and albedo for eight types of ecoroof soils with different moisture levels, commonly used for green roofs, are available in [29]. A correlation between thermal conductivity and thermal capacity for different kind of soils is presented in [30]. The soil used in the green roofs has contain a modest percentage of organic compost, relevant percentage of sand as well as light material such as pumice or expanded shale.

The canopy is constituted by leaves and air enclosed in the leaf cover, which is characterized by the following parameters:

- Height of plants;
- Leaf area index (LAI) that is the total one-sided area of leaf tissue per unit of ground surface area ($m^2 \cdot m^{-2}$);
- Minimum stomatal resistance, that governs the flow of water vapor through the stomates;
- Emissivity, reflectivity, absorbance and transmissivity of the leaves.

The LAI index ranges from 1.0 for sparse vegetation, to 5.0 for dense vegetation of grasses and small plants. This parameter characterizes the canopy–atmosphere interface, where most of the energy fluxes is exchanged [13] (Fig. 2.2).

Various studies have analyzed the thermal performance and the heat exchange phenomena which occur in green roofs [15, 31].

The soil acts as an inertial mass with a high thermal capacity and low thermal transmittance; the foliage behaves as a shading device under which convection provokes heat thermal exchange, but foliage absorbs part of the thermal energy for its vital process of photosynthesis; soil and vegetative layers induce evaporative and evapotranspiration cooling.

Incoming solar radiation is removed by sensible (convection) and latent (evaporative) heat flux from soil and plant surfaces, combined with conduction of heat into the soil substrate and long-wave (thermal) radiation [32, 33].

The thermal balance of extensive green roof indicates that heat fluxes are exchanged of about 50% by evapotranspiration, 30.9% by the long-wave radiative thermal flux and 1.2% accumulated and transferred to the rooms underneath [31, 34]. Green roofs reflect between 20 and 30% of solar radiation, and absorb up to 60% of it through photosynthesis, whereas a percentage below 20% of the heat is transmitted to the growing medium [23].

Fig. 2.2 Sparse (**a**) and large vegetation (**b**) [23]

2.3   Green Roof Modeling

The energy balance of the canopy is governed by radiative and convective thermal fluxes, evapotranspiration and evaporation/condensation of water vapor. Radiative thermal fluxes include the solar radiation absorbed by the leaves, the long-wave radiative exchanged between the leaves and sky, between the leaves and the soil surface and between the leaves themselves. Convective thermal fluxes occur between the leaves and the canopy air and between the soil surface and the canopy air. Evapo-transpiration includes the water evaporation into the leaves, the vapor diffusion onto the leaves surface and the vapor transport from leaves surface to the air.

The fast all season soil strength (FASST) model developed by the US Army Corps of Engineers is one of the numerical models most applied for simulating the energy balance of the green roofs [35]. The computational model includes: simplified moisture balance, soil and plant canopy energy balance, soil surface T_g and foliage T_f temperatures equations set are solved simultaneously each time step (Fig. 2.3).

The energy balance is divided into a budget for the foliage (1) and a budget for the ground surface (4) [34]. The energy balance on the foliage is based on the following equations:

$$F_f = \sigma_f \left[I_s^{\downarrow}(1 - r_f) + \varepsilon_f I_{IR}^{\downarrow} - \sigma \varepsilon_f T_f^4 \right] + \frac{\sigma_f \varepsilon_f \varepsilon_g \sigma}{\varepsilon_1} (T_g^4 - T_f^4) + H_f + L_f \qquad (2.1)$$

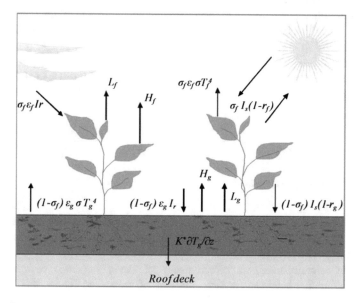

Fig. 2.3 Energy balance of the green roof

$$H_f = \left(1.1 LAI \rho_{af} C_{p,a} C_f W_{af}\right) \cdot \left(T_{af} - T_f\right) \tag{2.2}$$

$$L_f = l_f LAI \rho_{a,f} C_{p,a} C_f W_{af} r''(q_{af} - q_{f,sat}) \tag{2.3}$$

Equation (2.2) calculates the sensible heat flux between the leaf surface and air canopy, which depends by the temperature difference between leaf surfaces and canopy, the wind speed as well as the LAI.

Equation (2.3) provides the latent heat flux in the foliage layer, which depends on the transpiration phenomena of the leaves. It is mainly influenced by the stomatal and aero-dynamic resistance of the leaves, soil moisture content, light intensity and vapor pressure difference between the inside of the leaf and the outdoor atmospheric. The energy balance on the ground surface, is characterized by the following equations:

$$F_g = (1 - \sigma_f) \left[I_s^\downarrow (1 - r_g) + \varepsilon_g I_{IR}^\downarrow - \sigma \varepsilon_g T_g^4 \right]$$
$$- \frac{\sigma_f \varepsilon_f \varepsilon_g \sigma}{\varepsilon_1} \left(T_g^4 - T_f^4 \right) + H_g + L_g + K \frac{dT_g}{dz} \tag{2.4}$$

$$H_g = \rho_{ag} c_{p,a} C_{hg} W_{af}(T_{af} - T_g) \tag{2.5}$$

$$L_g = C_{e,g} l_g W_{af} \rho_{a,g}(q_{af} - q_{g,sat}) \tag{2.6}$$

The energy budget on the ground surface (2.4) is mainly influenced by the moisture content, thermal properties of the soil and amount of surface covered of the leafs. The sensible heat flux in the soil layer is given by the Eq. (2.5). It is function of the temperature difference between the soil surface and the air and by the wind speed of the air canopy. Equation (2.6) gives the heat latent exchanged between the ground surface and the air which depends on the difference between the mixing ratio of the soil surface and air and the wind speed of the air within canopy. The end result is a set of two simultaneous equations for temperatures, one for soil and the other for the foliage.

2.4 Methodology

2.4.1 Building Simulations

Green roofs are frequently designated as an efficient technologies that reduce the direct influence of solar radiations, indoor temperature fluctuations, and indoor air temperature peaks [15–18, 36].

The investigations of the effectiveness of the green roof to achieve the above-mentioned objectives requires the use of numerical model that allows to simulate the influence of the radiative properties and thermal inertia on the energy performance of building.

Nowadays, many software tools are available for building dynamic simulations [22]. In this study the software Design Builder has been used. In Design Builder the green roof module can be implemented as outer layer of a rooftop construction using a "Material: Roof Vegetation". The calculation method consists in a energy budget analysis that it follows the Fast All Season Soil Strength (FASST) model developed by Frankenstein and Koenig for the US Army Corps of Engineers [34].

This tool carries out accurate thermal analyses and allows very detailed inputs, including: climatic data (including air temperature, solar radiation, relative humidity hourly profiles); construction materials and components in dedicated libraries or manually edited; energy systems' specifications; time schedules (systems' management, occupancy, electric lighting, ventilation, etc.).

2.4.1.1 Thermal Dynamic Behavior

The dynamic thermal simulations permits to calculate the energy needs of both in heating and cooling period, the indoor air temperature (T_a), the outdoor (T_{so}) and in-door superficial temperature (T_{si}) and the indoor operative temperature.

The knowledge of these temperatures provides a set of data relating with the performances of the building envelope as thermal inertia parameters, indoor thermal comfort and the mitigation effects on the UHI [37].

As well known building energy performance strongly depends by the thermal inertia of the building envelope components. Buildings envelope with high thermal mass and heat capacity contribute to the reduction and the time delay of the peaks of cooling load [38].

The time lag (TL) and decrement factor (DF) are the two parameters widely used for characterizing the thermal inertia and heat storage capability of buildings.

The Time lag (TL) is measured by the delay of time between the hour when the maximum outer surface temperature occurs and the hour when the maximum inner surface temperature occurs.

$$TL = \tau_{T_{so,max}} - \tau_{T_{si,max}} \qquad (2.7)$$

The decrement factor (DF) is the ratio of the amplitude of the inner surface temperature fluctuations to the amplitude of the outer surface temperature fluctuations:

$$DF = \frac{A_{si}}{A_{so}} = \frac{T_{si,max} - T_{si,min}}{T_{so,max} - T_{so,min}} \qquad (2.8)$$

Therefore, it is worth to investigate the thermal behavior of the Green Roof with reference the above mentioned thermal inertia parameters.

Further, the temperature reached on the external building surface strongly influences the sensible heat available for transmission to the air or to building envelopes and, consequently the air temperature within the urban area.

2.4.1.2 Thermal Comfort

The purpose of ASHRAE Standard 55, thermal environmental conditions for human occupancy, is "to specify the combinations of indoor space environment and personal factors that will produce thermal environmental conditions acceptable to 80% or more of the occupants within a space" [39].

Actual thermal comfort is dependent on environmental factors, such as air temperature, air velocity, relative humidity and the uniformity of conditions, as well as personal factors such as clothing, metabolic heat, acclimatization, state of health, expectations, and even access to food and drink [40].

When calculating the thermal indoor climate, the operative temperature can be used as a simple measure for the heat loss from a person. It is the combined effects of the mean radiant temperature (T_{MRT}) and air temperature (T_{dry}).

Mathematically, operative temperature (T_{op}) can be expressed as;

$$T_{op} = (h_r \cdot T_{MRT} + h_c \cdot T_{db})/(h_r + h_c) \qquad (2.9)$$

where, T_{MRT} is the mean radiant temperature for the thermal zone, T_{db} is the mean zone air temperature, h_c is convective heat transfer coefficient; h_r is the linear radiative heat transfer coefficient.

The operative temperature (T_{op}) can also be calculated using the Eq. (2.10) as per Standard 55-2004.

$$T_{op} = \gamma \cdot T_{db} + (1 - \gamma) \cdot T_{MRT} \qquad (2.10)$$

γ is the radiant fraction, whose typical value is 0.5.

It can be useful in assessing the likely thermal comfort of the occupants of a building. Normally, whenever T_{op} exceeds desired comfort range, people tries to change the T_{db} set point to achieve a comfortable value of the operative temperature.

2.5 Pilot Study

2.5.1 Reference Building

The reference building is a typical holiday houses situated in the south coast of Sicily (lat. 36°53′, long. 4°25′). This typology of buildings (Fig. 2.4) were built at

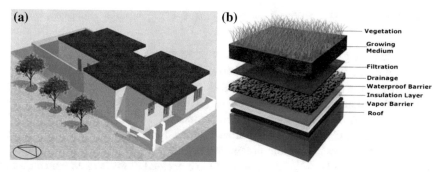

Fig. 2.4 Simulated building **a** 3D view, **b** green roof

least 30 years ago, when thermal insulation of the building envelope was almost totally disregarded in Italy.

The roof is a flat roof in reinforced concrete slabs with thickness of 25 (cm), the opaque vertical closures are characterized by masonries in double brick walls with internal air gap with an overall thickness of 30 (cm). Basic horizontal closure is composed by reinforced concrete slab and upper cement screed, coated with ceramic pavement.

The building is equipped with an air conditioning system, which operates with a set-point temperature of 20 °C during the heating period (December 1st–March 31th) and a set-point temperature of 26 °C during the cooling period (June 1st–September 30th).

As concerns internal gains, electric equipment and lighting: the average occupancy density is 0.04 people·m^{-2}, the density power for lighting and domestic equipment is 10 W·m^{-2} for the entire environment except the kitchen zone where the internal gains are 45 W·m^{-2}. The air change rate is 0.5 vol/h. The main geometrical feature of this building are reported in Table 2.1.

The thermal transmittance and mass surface of the building components are reported in Table 2.2.

Table 2.1 Geometric feature of the building and characteristic parameters

V (m^3)	S (m^2)	S/V (m^{-1})	S_u (m^2)	S_r (m^2)	h (m)
513	530	1.03	151.17	151.17	3.00

Table 2.2 Thermal transmittance and mass surface of building components

Building components	U (W m^{-2} K^{-1})	SM (kg m^{-2})
External masonry in double brick walls (30 cm)	1.27	381
Roof slab	3.17	571
Ground floor	3.77	474
Windows	2.60	–

2.6 Building Retrofits Scenario

A suitable strategy, to improve the energy performance of this typology of buildings, is the refurbishment of the roof considering that usually roofs have a large dispersing surface. Therefore, three scenario have been investigated, two cases foresee the addition of thermal insulation, with a thickness of 4.0 and 8.0 cm respectively, the third case foresee the implementation of an extensive green roof.

It has been proven that non-insulated green roofs can be advantageous in summer, but they less favorable in winter when the roof would benefit of a higher thermal resistance [41, 42]. Therefore, we have decided to investigate a green roof averagely insulated with 4.0 cm of thermal insulation, as an efficient way to obtain good performance of both during warmer days and, during colder ones.

2.6.1 Descriptions of the Green Roof

An extensive green roof has been considered because it is compatible with the existing building structures. The vegetation layer can be constituted by mosses, sedum, graminaceous or other succulents plants, which are very common and suitable plants for using on an extensive green roof. All the above mentioned plants are small plants that grow across the ground rather that upwards offering good coverage and roof membrane protection [18]. Moreover such plants have the capabilities to store water in their leaves making them high drought resistance. The thermo-physical properties of vegetation layer and substrate are reported in Table 2.3.

The substrate is a thin layer (10 cm) of porous soil, it is typically a mixture of sand, clay, mineral aggregates and organic matter. The soil is above the filter layer, a geotextile fabric, that filter the soil granules in order prevent that the empties of the drainage layer are filled. Table 2.4 summarizes the main characteristics of the proposed green roof.

Table 2.3 Thermo-physical properties of vegetation layer

Vegetation layer			Substrate		
Height of the plants	H	0.10 (m)	Thermal conductivity	λ	0.98 (W m^{-1}K^{-1})
Leaf area index	LAI	5.0 (m^{-2} m^{-2})	Density	ρ_g	1460 (kg m^{-3})
Reflectivity	r_f	0.22	Specific heat	C_p	880 (J kg^{-1} K^{-1})
Absorbance	α_f	0.60	Emissivity	ε_g	0.90
Transmissivity	t_f	0.18	Absorbance	α_g	0.70
Emissivity	ε_f	0.95	Residual moisture content	θ_r	0.01 (m^{-3} m^{-3})
Minimum stomatal resistance	r	180 (s m^{-1})	Initial moisture content	θ_{in}	0.15 (m^{-3} m^{-3})

Table 2.4 Stratigraphy and thermal properties of the extensive green roof

Layers	s (cm)	λ (W m^{-1} K^{-1})	ρ (kg m^{-3})	C_p (J kg^{-1} K^{-1})
1. Vegetation layer	–	–	–	–
2. Soil layer	10.0	0.98	1460	880
3. Filter layer	0.1	0.22	910	1800
4. Drainage layer	10.0	0.92	900	1000
5. Root barrier	0.5	0.19	1400	1200
6. Waterproof membrane	0.5	0.19	1400	1200
7. Thermal insulation	0.4	0.033	100	710
8. Moisture barrier	0.5	0.23	1100	1050
9. Concrete slab	24.0	2.30	2300	1000

The U-value of this extensive green roof results of 0.55 W m^{-2} K^{-1} and its surface mass (SM) is 762 kg m^{-2}.

2.7 Energy Performance Simulations

2.7.1 Building Simulations

The climatic conditions, air temperature and relative humidity as well as the solar radiation, are taken by the wheatear database embedded in Energy Plus for the Catania [22]. Figure 2.5 depicts the average daily solar irradiation on an horizontal plane and the outdoor air temperature.

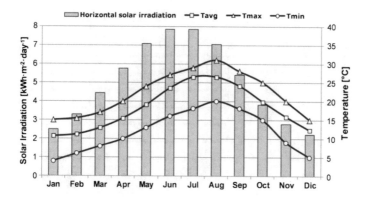

Fig. 2.5 Main average weather data of Catania city

Table 2.5 Thermal transmittance and mass surface of the roofs considered

Roofs	Acronym	Thickness (m)	Insulation thickness (m)	U-value (W m^{-2} K^{-1})	SM (kg m^{-2})
Existing baseline roof	TR$_0$	0.29	0	3.17	645
Averagely insulated roof	TR$_4$	0.34	0.04	0.61	653
Well insulated roof	TR$_L$	0.38	0.08	0.35	657
Averagely insulated green roof	GR	0.55	0.04	0.55	762

The calculation of the overall building energy performance were carried out from June 1st to September 30th (cooling period) and from December 1st and March 31th (heating period).

The energy performance of the green roof scenario has been compared with the performance of the existing baseline roof and with the other two insulated roof, averagely and well insulated. The main termo-physical characteristics of all the investigated roofs scenario are reported in Table 2.5.

Figure 2.6 shows the comparisons of the heat fluxes exchanged through the roof surfaces for each investigated scenario.

The GR achieves the better results in the cooling period, because the incoming heat flux through the GR is only the 5% than existing baseline roof. The well insulated roof TR$_L$ achieves the best performances in heating period with a reduction of the of heat flux of (83%) compared with the bare roof. During the heating period, it can be noted that the GR achieve performance that are a slightly lower than the well-insulated roof TR$_L$, which has 4 cm of extra thermal insulation.

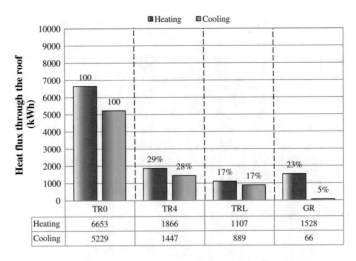

	Heating	Cooling
Heating	6653	1866	1107	1528
Cooling	5229	1447	889	66

Fig. 2.6 Heat flux exchanged through the roof in heating and cooling period

2.7.2 Energy Needs

The energy needs of the building were calculated of both in the heating and cooling period. The results comparison of the yearly specific energy needs (kWh/m²y) are reported in Fig. 2.7a, b.

As concerns the cooling period, it can be noticed that the reductions of the energy needs achievable with the green roof scenario (GR) are of about 90, 76 and 69%.

When compared respectively with the existing baseline (TR_0), the intermediate and well insulated roofs. As concerns the heating period (from 1st December to 31st March) it is can be noticed that the well-insulated roof allows to achieve better performance than the green roof scenario, while the GR and the intermediate insulated roofs (TR_4) achieve about the same performance. However, the energy deficit between the GR scenario and the well-insulated scenario is just about 3%. The specific energy needs for heating and cooling of the building are summarized in Fig. 2.8.

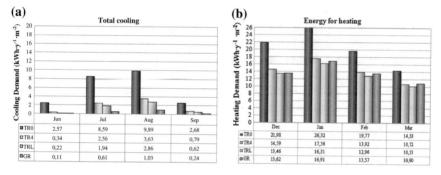

Fig. 2.7 Specific energy needs of each roof scenario: **a** cooling, **b** heating

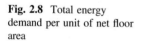
Fig. 2.8 Total energy demand per unit of net floor area

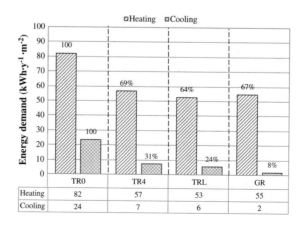

Finally, it is possible to assert that the GR scenario provides the best performance considering an all year round analysis.

Moreover, the reduction of the energy needs in summer period implies the decreasing of the heat discharged through the condenser of the AC plants in the urban area. This aspect surely contributes to mitigate the UHI intensity at micro scale.

2.7.3 Assessment of the Thermal Dynamic Behaviour

With the aim to analyses the effectiveness of the GR to reduce indoor and outdoor superficial temperature fluctuations the hourly and daily profiles of both the superficial, internal (T_{si}) and external (T_{so}) as well as indoor temperature (T_i) were calculated in free running conditions.

The daily temperature profiles of the baseline roof, the well-insulated and the green roofs scenario are reported in Fig. 2.9, during the period (15st July–15th August).

The temperatures of the averagely insulated scenario is not reported, since its thermal dynamic behavior lies between the TR_0 and the TR_L temperature profiles.

It is possible to notice that throughout the period investigated the outdoor superficial temperature of the baseline $T_{so}(TR_0)$ and the well-insulated roof $T_{so}(TR_L)$ are always higher than and $T_{so}(GR)$. The average monthly difference between $T_{so}(GR)$ and $T_{so}(TR_L)$ is of about 7.0 °C, with a maximum daily value of 9.5 °C. Whereas the value of $T_{si}(TR_L)$ and $T_{si}(GR)$ are almost comparable with mean differences of about 1.0 °C. These results indicate the barrier effect of the

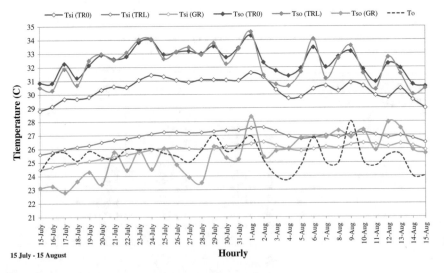

Fig. 2.9 Daily profiles of outer and inner surface temperature

thermal insulations, which allow a very large increase of the superficial external temperature.

These results indicates the very appreciable action of the green roof for mitigate the UHI, thanks to its lowest superficial outdoor temperatures.

Indeed, the lowest temperature on the external surface of the GR reduce the transmission of sensible heat flux from the building to the urban environment.

2.7.3.1   Assessment of the Thermal Dynamic Behaviour

In addition, the time lag and the decrement factors have been calculated for describing the response of the green roof to outer forcing conditions, using the Eqs. 2.7 and 2.8.

Figure 2.10 shows the comparison between the hourly variations of the superficial temperatures, in free running conditions, of the GR, TR_0 and TR_L scenario in a day characterized by the highest peaks of outdoor temperature (10 August).

It is evident the very wide temperature fluctuations of both TR_0 and TR_L when they are compared with the GR. Table 2.5 summarize the main results (Table 2.6).

It is possible to notice that the green roof achieves the lowest peak of outdoor surface temperature ($T_{so,max} = 32$ °C), that is 13 and 16 °C lower than the other two scenario. On the other hand, the well-insulated roof is characterized by the highest value of peak of the outdoor surface temperature ($T_{so,max} = 48$ °C).

The well-insulated roof, TR_L, and the green roof scenario, GR, have almost similar values of both time lag and decrement factor.

The well-insulated roof reaches such high value of $T_{so,max}$ because the insulating layer stops the heat flux from the outdoor to indoor causing the overheating of the

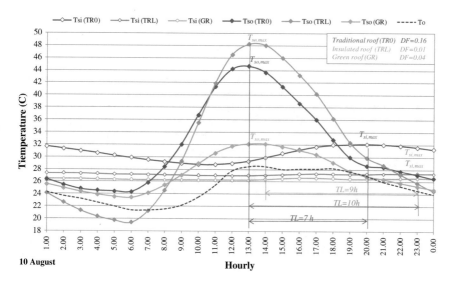

Fig. 2.10 Hourly profiles of the outer and inner surface temperature

Table 2.6 Dynamic thermal parameters

Roof	$T_{so,max}$ (°C)	$T_{si,max}$ (°C)	TL (h)	DF (-)
TR_0	45 °C (13:00)	32 °C (20:00)	7.0	0.16
TR_L	48 °C (13:00)	27 °C (23:00)	10.0	0.01
GR	32 °C (14:00)	26 °C (23:00)	9.0	0.04

outer surface of the roof and, consequently contribute to increase the local urban warming.

Thereby the main advantage of the GR in comparison with TR_L is its capability to maintain lower external temperatures.

2.7.4 Assessment of Thermal Comfort

This section assesses the effect on the indoor thermal comfort conditions induced by the green roof in comparison with the other roof scenarios considered. The comparison has been performed using the temperature (T_{op}) calculated using Eq. (9).

Figure 2.11 shows the hourly profiles of T_{op} calculated for each scenario during three typical summer days, from 8 to 10 August.

Subsequently, considering the value of 26 °C as a setpoint of the operative temperature, the percentage of hours when T_{op} is higher than 26 °C has been carried out.

Table 2.7 reports, the maximum and the average values of the operative temperatures calculated for each scenario during the investigated period and, the corresponding predicted mean vote (PMV) as well as the percentage of dissatisfied (PPD).

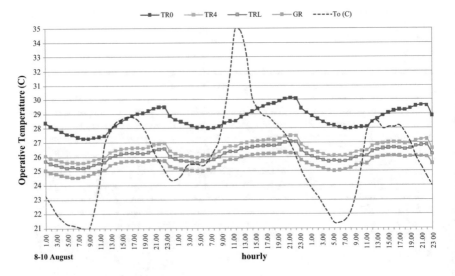

Fig. 2.11 Hourly profiles of the operative temperature

Table 2.7 Average operative temperatures (8–10 August)

Values	TR_0	TR_4	TR_L	GR
$T_{op,max}$ (°C)	30.13	27.50	27.08	26.33
PMV/PPD	1.59/56%	0.70/15%	0.56/12%	0.31/7%
$T_{op,av}$ (°C)	28.65	26.53	26.15	25.53
PMV/PPD	1.09/30%	0.38/8%	0.25/6%	0.04/5%
$T_{op,av} > 26$ °C	100%	83%	65%	26%

The PMV was calculated considering the following assumptions: air speed 0.1 m/s; humidity 50%; Metabolic rate 1.1 met; clothing rate 0.5 cloth [43].

Further, the percentage of hours when $T_{op} > 26$ °C is also shown.

It is quite evident that the scenario TR_0 involves the worst performances with thermal sensation between Slightly Warm and Warm when the average or the max values of T_{op} are used.

The intermediate and well-insulated scenarios provide thermal sensations of Slightly Warm or neutral in conjunction with the max or average values of T_{op}.

Otherwise the GR scenario leads to the better results, indeed the neutral thermal sensation is always guaranteed and it complies with ASHRAE Standard 55-2013.

Figure 2.12 shows in psychometric chart the comfort zone boundary and the thermodynamic state of the GR scenario with the $T_{op,av}$.

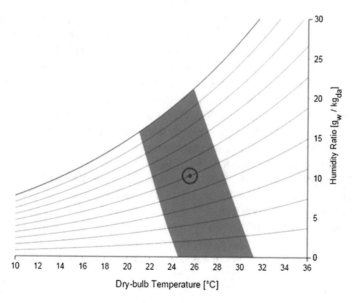

Fig. 2.12 Comfort zone and thermal comfort for GR scenario

2.8   Discussion

The results of this research are in accord with other literature studies that indicate an energy savings for building cooling of a nursery school in Hong Kong of about 75% [32], and of about 70% for a hospital building sites in Vicenza (Italy) [33]. Otherwise, an extensive green roof applied on a nursery school in Athens (Greece) provides a reduction of the building cooling load of about 50% [37].

The dynamic simulations also have proven the reduction of the absolute superficial temperature and their daily fluctuations from 20 to 6 °C, in comparison with the bare roof scenario. This result agree with the reduction of temperature fluctuations calculated by [42] and [44] which have found a temperature reduction from 18 to 6 °C.

The effect of the dynamic properties was investigated for a building sited in Floria-nopolis [14], which calculated a (TL) of about 10 h, that is really comparable with the (TL) of 9 h calculated in this study. The increase of the delay time is also confirmed by experimental study conducted [45].

2.9   Conclusions

The research presented in this paper, investigates the thermal and energy performance of an existing buildings on which it was applied an extensive green roof.

The thermal behavior of such building, has been investigated through dynamic simulations for assessing the energy needs, the variations of the superficial temperature and the indoor thermal comfort. Moreover, other two scenario that foresee the addition of two different thickness of thermal insulation, intermediate and well-insulated, have been investigated and compared with the green roof scenario.

The discussion of the results highlights that the green roof allows to obtain the highest energy saving in summer. The energy savings obtainable are of about 92% and 33% in the cooling and winter period, respect to the bare roofs. Generally, the GR scenario allows to obtain better performances in comparison with the bare roof (TR_0) and the averagely insulated roof (TR_4).

Instead the green and the well-insulated roof (TR_L) allow to achieve comparable energy savings, quite similar thermal inertia parameters, time lag and decrement factors, and to achieve good indoor thermal comfort.

The main advantage of the GR in comparison with TR_L is its capability to maintain lower external temperatures. Indeed differences of the peak values of the outdoor surface of 16 °C between these two scenario have been found.

The well-insulated roof reaches such high value of $T_{so,max}$ because the insulating layer stops the heat flux from the outdoor to indoor causing the overheating of the outer surface of the roof. Thereby, it is possible to highlight that well-insulated roofs allow to achieve substantial energy savings, but on the other hand they act as a catalyser that contribute to increase the local global warming.

As results, the lowest temperature on the external surface of the GR reduces the transmission of sensible heat flux from the building to the urban environment and, obviously contribute to mitigate the UHI effects. Therefore, the benefits of green roof as countermeasure for mitigating the UHI have to include the low outdoor surface temperature and the energy savings in summer period this last benefit reduces the heat discharged by the condenser of the AC plants in the crowded urban environment.

Thus, the energy efficiency improvements of existing buildings may be considered as part of a strategy to pursue and expand the scope of climate change adaptation measures.

References

1. Theodosiou, T.G.: Summer period analysis of the performance of a planted roof as cooling passive technique. Energy Build. **35**(9), 909–917 (2003)
2. D'Orazio, M., Di Perna, C., Di Giuseppe, E.: The effects of roof covering on the thermal performance of highly insulated roofs in Mediterranean climates. Energy Build. **42**, 1619–1627 (2010)
3. Gagliano, A., Patania, F., Nocera, F., Ferlito, A., Galesi, A.: Thermal performance of venti-lated roofs during summer period. Energy Build. **49**, 611–618 (2012)
4. GhaffarianHoseini, A., Dahlan, N., Berardi, U., Makaremi, N.: Sustainable energy performances of green buildings: A review of current theories, implementations and challenges. Renew. Sustain. Energy Rev. **25**, 1–17 (2013)
5. Coutts, A., Daly, E., Beringer, J., Tapper, N.J.: Assessing practical measures to reduce-urban heat: Green and cool roofs. Build. Environ. **70**, 266–276 (2013)
6. Alexandri, E., Jones, P.: Temperature decreases in an urban canyon due to greenwalls and green roofs in diverse climates. Build. Environ. **43**(4), 480–493 (2008)
7. Castleton, H.F., Stovin, V., Beck, S.B.M., Davison, J.B.: Green roofs; building energy savings and potential for retrofit. Energy Build. **42**, 1582–1591 (2010)
8. Mentens, J., Raes, D., Hermy, M.: Green roof as a tool for solving the rainwater runoff problem in the urbanized 21st century? Landscape Urban Plann. **77**, 217–226 (2006)
9. Fioretti, R., Palla, A., Lanza, L.G., Principi, P.: Green roof energy and water related performance in the Mediterranean climate. Build. Environ. **45**, 1890–1904 (2010)
10. Berndtsson, J.C., et al.: Runoff water quality from intensive and extensive vegetated roofs. Ecol. Eng. **35**(3), 369–380 (2009)
11. Yang, J., Yu, Q., Gong, P.: Quantifying air pollution removal by green roofs in Chicago. Atmos. Environ. **42**(31), 7266–7273 (2008)
12. Banting, D., et al.: Report on environmental benefits and costs of green roof technologies for the city of Toronto (2005)
13. Susca, T., Gaffin, S.R., Dell'Osso, G.R.: Positive effects of vegetation: Urban heat island and green roofs. Environ. Pollut. **159**, 2119–2126 (2011)
14. Parizotto, S., Lamberts, R.: Investigations of green roof thermal performance intemperature climate: A case study of an experimental building in Florianopolis city, Southern Brazil. Energy Build. **43**, 1712–1722 (2011)
15. Del Barrio, E.P.: Analysis of the green roofs cooling potential in buildings. Energy Build. **27**, 179–193 (1998)
16. Niachou, K., Papakonstantinou, K., Santamouris, A., Tsagroussoulis, A., Mihalakakou, G.: Analysis of the green roof thermal properties and investigations of its energy performance. Energy Build. **33**, 719–729 (2001)

17. Lui, K., Minor, J.: Performance evaluation of an extensive green roof. Greening Rooftops for Sustainable Communities, pp. 1–11. Washington DC (2005)
18. Kumar, R., Kaushik, S.C.: Performance evaluation of green roof and shading for thermal protection of buildings. Build. Environ. **40**(11), 1505–1511 (2005)
19. Costanzo, V., Evola, G., Gagliano, A., Marletta, L., Nocera, F.: Study on the application of cool paintings for the passive cooling of existing buildings in mediterranean climates. Adv. Mech. Eng. (2013)
20. Teemusk, A., Mander, U.: Green roof potential to reduce temperatures fluctuations of a roof membrane: A case study from Estonia. Build. Environ. **44**(3), 643–650 (2009)
21. Takakura, T., Kitade, S., Goto, E.: Cooling effect of greenery cover on over a building. Energy Build. **31**(1), 1–6 (2000)
22. DesignBuilder—Energy Simulation Software, Version 3. http://designbuilder.co.uk
23. La Roche, P., Berardi, U.: Comfort and energy savings with active green roofs. Energy Build. **82**, 492–504 (2014)
24. Capelli, M., Cianfrini, C., Corcicone, M.: Effect of vegetation roof on indoor temperatures. Heat Environ. **16**(2), 85–90 (1998)
25. Karachaliou, P., Santamouris, M.: Experimental and theoretical analysis of the thermal behavior and the energy performance of an intensive green roof system installed on an office building in Athens. In: Proceedings of the Third International Conference on Countermeasures to Urban Heat Island Venice (2014)
26. Gagliano, A., Detommaso, M., Nocera, F., Patania, F., Aneli, S.: The retrofit of existing buildings through the exploitation of the green roofs–a simulation study. Energy Procedia **62**, 52–61 (2014)
27. http://www.zinco-greenroof.com/EN/downloads/pdfs/ZinCo_Intensive_Green_Roofs.pdf
28. Farouki, O.T.: Thermal properties of soils. U.S. Army Cold Regions Research and Engineering Laboratory, Monograph 81-1 (1981)
29. Sailor, D.J., Hutchinson, D., Bokovoy, L.: Thermal property measurements for ecoroof soils common in the western U.S. Energy Build. **40**, 1246–1251 (2008)
30. Capozzoli, A., Gorrino, A., Corrado, V.: Thermal characterization of green roofs through dynamic simulation. In: Proceedings of BS2013 13th Conference of International Building Performance Simulation Association, Chambéry (2013)
31. Feng, C., Meng, Q., Zhang, Y.: Theoretical and experimental analysis of the energy balance of extensive green roofs. Energy Build. **42**(6), 959–965 (2010)
32. Wong, N.: Investigation of thermal benefits of rooftop garden in the tropical environment. Build. Environ. **38**(2), 261270 (2003)
33. Lazzarin, R.M., Castellotti, F., Busato, F.: Experimental measurements and numerical modeling of a green roof. Energy Build. **37**(12), 1260–1267 (2005)
34. Energy Plus.: Engineering Reference. The reference to energy plus calculation, Green Roof Model (EcoRoof). University of Illinois and University of California, pp. 123–132 (2011)
35. Frankenstein, S., Koenig, G.: Fast all-season soil strength (FASST), U.S. Army Engineer Research and Development Center, Colde Regions Research and Engineering Laboratory (ERDC/CRREL), Special Report SR-04-01. Washington, DC
36. Santamouris, M., Pavlou, C., Doukas, P., Mihalakakou, G., Synnefa A., Hatzibiros, A.: Investigating and analysing the energy and environmental performance of an experimental green roof system installed in a nursery school building in Athens, Greece. Energy. **32**, 1781–1788 (2007)
37. Li, D., Bou-Zeid, E ., Oppenheimer, M.: The effectiveness of cool and green roofs as urban heat island mitigation strategies. Environ. Res. Lett. **9** (2014). doi:10.1088/1748-9326/9/5/055002
38. Gagliano, A., Patania, F., Nocera, F., Signorello, C.: Assessment of the dynamic thermal performance of a massive building. Energy Build. **72**, 361–370 (2014)
39. ASHRAE Standard 55—thermal environmental conditions for human occupancy, ASHRAE Inc., (2004)

40. Tina, G.M., Gagliano, A., Nocera, F., Patania, F.: Photovoltaic glazing: Analysis of thermal behavior and indoor comfort. Energy Procedia **42**, 367–376 (2013)
41. La Roche, P.: Low cost green roofs for cooling: experimental series in a hot anddry climate. In: Passive Low Energy Conference, PLEA 2009, Quebec, Canada (2009)
42. Zinzi, M., Agnoli, S.: Cool and green roofs. An energy and comfort comparison between passive cooling and mitigation urban heat island techniques for residential buildings in the Mediterranean region. Energy Build. **55**, 66–76 (2012)
43. http://smap.cbe.berkeley.edu/comforttool/
44. Eumorfopoulou, E., Aravantinos, D.: The contribution of a planted roof to the thermal protection of building in Greece. Energy Build. **27**, 29–36 (2003)
45. D'Orazio, M., Di Perna, C., Di Giuseppe, E.: Green roof yearly performance: A case study in a highly insulated building under temperate climate. Energy Build. **55**, 439–451 (2012)

Chapter 3
Understanding Opportunities and Barriers for Social Occupant Learning in Low Carbon Housing

Magdalena Baborska-Narozny, Fionn Stevenson and Paul Chatterton

Abstract The effectiveness of the collective learning that takes place in modern housing developments can play a major role in terms of housing performance. Building performance evaluation (BPE) currently does not address the type and quality of collective learning processes happening within a community in relation to occupants using their new homes. A Social Learning Tool is proposed to extend BPE methodology and provide a framework to help researchers better understand the nature and degree of home user collective learning and community involvement which can in turn enhance the BPE process. A first partial application of the tool to six case study dwellings within a low carbon development in Leeds allowed identification of barriers and opportunities for collective learning.

Keywords Collective learning · Home handover process · Home use · Building performance evaluation · Mechanical ventilation with heat recovery

M. Baborska-Narozny (✉)
Faculty of Architecture, Wroclaw University of Science and Technology, Wroclaw, Poland
e-mail: magdalena.baborska-narozny@pwr.edu.pl

F. Stevenson
School of Architecture, University of Sheffield, Sheffield, UK
e-mail: f.stevenson@sheffield.ac.uk

P. Chatterton
School of Geography, University of Leeds, Leeds, UK
e-mail: p.chatterton@leeds.ac.uk

© Springer International Publishing AG 2017
J. Littlewood et al. (eds.), *Smart Energy Control Systems for Sustainable Buildings*,
Smart Innovation, Systems and Technologies 67,
DOI 10.1007/978-3-319-52076-6_3

3.1 Introduction

As 'zero energy[1] performing houses become more and more technically feasible [1–3], the variation in energy use related to the inhabitants comes increasingly into focus [4–8]. This is in the context of the energy use reduction strategy for the UK housing, which is based on a 29% reduction on 2008 carbon dioxide emissions by 2020 [9]. Repeated studies have shown that variations between adopted habits and home use patterns are a major factor behind a 300% difference in energy performance observed between houses of exactly the same specification [10, 11]. As Gram-Hanssen points out "It is absolutely possible to be highly engaged in the environment without knowing technically how to influence the level of consumed energy, as it is also possible to be technically interested in doing 'the right thing' without being especially environmentally concerned..." ([10], p. 185) Gram-Hanssen's finding was based on analysis of heating practices of a group of residents with varied interest in environmental agenda who lived in a Danish development built between the 1960s–70s, which was equipped with relatively simple technology compared to recent low energy housing. An early assumption of the case study research project reported here was that in recent increasingly technically complex homes the inhabitants need to gain new technical knowledge to stay in control of the internal environment and achieve or exceed the target performance [12, 13]. These new skills and understanding go beyond the tacit knowledge built up over time in previous accommodation due to changes in technology installed. It was further assumed that inhabitants who hold positive environmental values would be actively involved in mastering their understanding and skills of efficient home use.

Efficient home use is defined here as achieving a healthy indoor environment and comfort while minimizing energy consumption. In the longer term it also needs to includes appropriate maintenance strategies, to provide a more comprehensive lifecycle approach. The pioneering 'low impact' ethos and 'best practice' ambitions of the case study development selected for the research suggested, according to planned behaviour theory [14], that increased peer pressure among residents would help minimise their homes' in-use energy performance. The social structure of this Co-housing type of the development also raised an expectation of collective home use learning taking place. The presumed technology related learning phase leading to daily practices formation was the major theme shaping the one year long field study of this new built UK housing development with its strong environmental agenda [15]. As one of the outcomes of this case study, this chapter focuses on identified opportunities and barriers to engage in home use learning related to technology and proposes a theoretical framework to guide future research.

---

[1]'Zero Energy' as proposed in article 2 of EU Directive 2010/31/EU (EPBD recast) that is including energy in use but not including embodied energy or complete energy lifecycle of construction and home components.

When occupants move into a new housing development they are all simultaneously faced with a similar set of challenges related to the development. However, their skills, time resources and motivation to tackle these challenges may vary substantially. To date, domestic building performance evaluation (BPE) has been largely concerned with the process of delivering buildings, the resulting performance and response of individual occupants [16]. Lessons learned from BPE are mostly addressed to building industry actors [17]. As a result the 'Soft Landings' process has been developed in the UK to help improve industry practices by closing the gaps in knowledge transfer between different stages and actors [18]. Research into social learning between inhabitants and professionals indicates a similar gap in knowledge transfer between these groups [19]. A corresponding 'Soft Landings' process aimed at facilitating home use skills and knowledge exchange among the residents and between residents and external networks is therefore also needed. The first step towards this is to better understand the dynamics of technology related learning process and social learning potential [20–22] which goes beyond the current learning model.

Home use learning is predominantly seen by the industry as an individual linear process beginning with a home demonstration tour, followed by the mastering of skills explained in a Home User's Guide (HUG) with a clear objective: to close the gap between the user's understanding and the design intention [23, 24]. This main-stream model fits with Badura's social learning theory that focuses on individual observational learning [25]. There are four major assumptions underpinning this process when applied to the home use learning context which are that:

- the design model, systems and controls provided actually match occupant needs
- all systems work well
- the learning support provided is correct and presented in a way that enhances users' understanding
- occupants engage in the learning process effectively.

Any failure in any of these areas can lead to a basic misunderstanding or preserve the user within a state of 'unknown-unknowns', where they literally do not know what they should know, leading to the formation of inefficient habits in home use.

3.2 Home Use Social Learning Conceptual Framework

One way to understand the term social learning is Sørensen's ([26] p. 6) useful definition: '[Social learning] can be characterised as a combined act of discovery and analysis, of understanding and giving meaning, and of tinkering and the development of routines. In order to make an artefact work, it has to be placed, spatially, temporally, and conceptually. It has to be fitted into the existing,

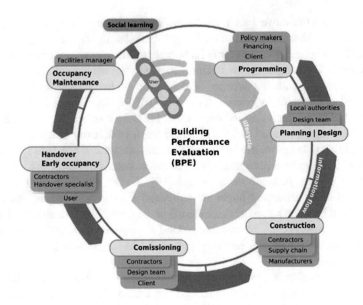

Fig. 3.1 BPE model enhanced with social learning element

heterogeneous networks of machines, systems, routines, and culture.' As Glad [19] points out, this relates to the following themes: user access to technology (physical and cognitive), communication (between users), trust (in technology and between users), social roles (of users), and co-production of technology. This chapter is largely focused on the user access to technology and technology related communication between inhabitants as critical points of learning intervention. The social learning (SL) analytical framework proposed here identifies components necessary to understand the challenge home users face and social interactions, opportunities and barriers for collective learning of efficient home use. It builds on a typical BPE process [27] and expands its scope from individual dwelling to a community level (Fig. 3.1).

The developed SL framework looks at the typical phase of home use learning during early occupancy but also considers the earlier stage when a residents group is first established and its identity and structure are defined (Fig. 3.2). The SL framework is built around a number of research questions identified as most relevant for judging whether collective home use learning would be beneficial and what form could it take within a specific development.

- What are home use expectations, prior experiences and skills?
- What is the home use learning challenge?
- What individual home use learning support is provided?
- What is the point of seeking home use improvement and is it clear to all involved?
- How are decisions made at a development level?

- Is there a variety in skills and understanding of home use within the development?

These questions set the scope of the extended BPE research proposed below and are answered for a case study example.

The home use learning challenge can be tackled through a combination of observational learning and development of new ideas unique for a specific context. Observational learning embraces skills and knowledge that already exist and can be acquired through observation and imitating behaviour. The term derives from social learning theory [25] which distinguishes three elements of the process: attention, retention and reproduction. The home demonstration tour is an explicit example of observational learning by the inhabitant. Its efficiency can be increased through securing a focused atmosphere (attention), hands on experience (reproduction) and a good HUG for future reference (retention) [23]. Long term focus (attention) on home use learning depends on awareness of the goals of the whole process. Awareness may be built through following and understanding one's own energy and water consumption and referring this to relevant benchmarks [5].

Social learning theory also acknowledges the role of motivation in the learning processes. In the SL framework, the motivation to learn about efficient home use may be both external, through social pressure, and internal, based on the occupant's own attitudes and expectations. Therefore occupants may be interested in improving home use practices to save natural resources, lower their own bills,

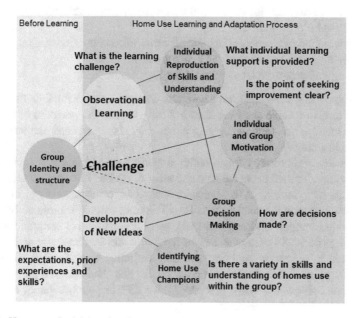

Fig. 3.2 Home use Social learning framework research questions

improve the internal environment and ease of achieving comfort, etc. Motivation is also directly linked with the decision making process within a community which is instrumental for effective implementation of both bottom-up and top-down initiatives within a development and enhancing information exchange between inhabitants—this is important to trigger collective home use learning. The decision making process adopted within a community determines each inhabitants long term role in governing and maintaining a development. The wide spectrum of such a role varies from a passive customer, with no power to influence maintenance service provided by an external organisation, to an active participant involved in shaping the maintenance of a development. System learning, experimentation, polycentricity and participation have all been identified by Biggs et al. as key attributes of a resilient governance model [28]. In a housing development context the resilience of its governance is tested whenever home use or maintenance issues arise and are not solved to residents' satisfaction. Resilience is lacking if such difficulties leave occupants with predominant feelings of frustration and anger. Resilience is strong when such identified weaknesses are seen as a challenge and they trigger process of improvements at development level. Collective home use learning is therefore probably more efficiently facilitated in a development with governance that has strong resilience, once a need for it is recognised, however this requires further investigation.

If collective learning is to become more than means for troubleshooting of home use issues the process needs to involve:

- Identifying areas for development through direct and developmental learning
- Seeking solutions: short and long term
- Introducing the most appropriate and efficient information flow within the group
- Identifying home use champions and supporting them
- Securing and storing lessons learned for future reference.

It is argued that home use information exchange between inhabitants which is only restricted to trouble shooting does not fully exploit the potential of collective home use learning as some underlying issues may stay unnoticed and solutions to these remain unexplored.

Home use learning may happen within a whole range of learning environments ranging from individual occupant trial and error or on-line information search, peer to peer learning, group meetings between occupants, or occupants and experts, or communication through social media. The SL framework maps these activities and aims to evaluate which environments work best for what type of community groups and learning needs. An understanding of the fabric and systems provided and their performance in homes is vital for researchers aiming to evaluate the home use learning stage, users' needs and particular challenges. The research guided by the SL framework should always be preceded with a BPE or post occupancy evaluation (POE) of a residential development.

3.3 Methods

To illustrate social learning framework application in an actual housing development context a case study methodology is utilized [29]. The Low Impact Living
affordable Community (LILAC) in Leeds, England was selected for the collective
home use learning study as it is an intentional community faced with a challenge of
'low impact living' in low energy housing. LILAC is a co-housing community,
established in 2006, with a vision of affordable low impact urban living and an
aspiration to trigger change in a wider social and economic context. LILAC has
always been engaged in a large variety of outreach activities on different levels to
promote values and solutions for future implementation by others. Its 35 adult
members with 9 children moved into a new innovative housing development of 20
dwellings in spring 2013. All the residents voluntarily develop and sign into the
goal of 'low impact living' and agree on common core values such as environmental sustainability and self-reliance [30]. The community's concern for the
environment has led to selection of prefabricated straw and timber technology [31],
a novelty in UK housing, and installing systems intended to diminish household's
energy use, mechanical ventilation with heat recovery (MVHR) among others.

The development comprises of two storey terraced houses and apartments in
three storey blocks and a communal house. All households agreed to participate in
the study. This involved analysis of the procurement process, fabric and mechanical
and electrical systems provided internal environment quality, user's interaction with
controls and varied community level aspects of the development (see Fig. 3.3 blue
dots). This ensured that the context of group activities, information flow within
community and decision making process were adequately mapped [32].

The research methods used in this case study can be broken down into two
categories: the methods used to establish baseline physical data and the qualitative
data checked against this baseline data.

3.3.1 Quantitative Monitoring

For the baseline data, each dwelling was equipped for one year with three discrete
sensors (I-buttons) monitoring dry bulb temperature and relative humidity (RH) in
different rooms with readings logged every half an hour and sensors placed at the
height of ca. 0.7 m above floor level away from direct sunlight. In the open plan
dining/kitchen/living area temperature only was monitored. In bathrooms and
bedrooms temperature and RH readings were taken. Additionally 3 selected
dwellings had CO_2 and RH data loggers installed in the dining area (Telaire CO_2
sensors and Hobo data loggers). Energy and water meter readings were regularly

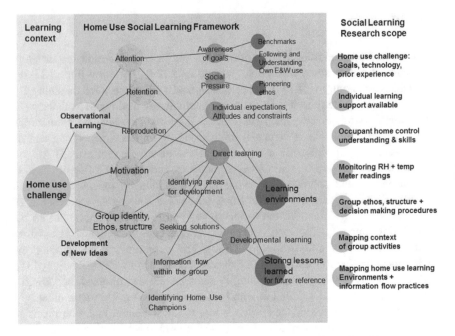

Fig. 3.3 Social learning tool conceptual framework showing scope covered in this paper

taken for each dwelling. A construction audit, and general services installation and commissioning checks were also performed. An air flow rate check was performed for MVHR system in 4 selected dwellings.

3.3.2 Qualitative Building and User Related Data

The temperature and RH sensors selected for the study required regular downloading visits every 8 weeks due to limited memory capacity. This sequence of regular home visits throughout the year, although time consuming, was a rich source of ethnographic observations and informal discussions on home use practices and mapping information flow practices within the community. Data obtained during these visits, including walk-through method, allowed critical understanding of results from more structured occupant feedback: two surveys described below and a 45 min semi-structured interview.

A 45 min client and design team interview was conducted to understand initial design intentions and procurement process. Associated occupant handover procedures were also scrutinised. The researcher shadowed twice an innovative handover procedure in LILAC. As a part of action research feedback was given covering guidelines of best practice handovers and a SWOT analysis of the observed process. Home user's guides were also analysed. The handovers and HUGs informed the starting point of home use learning process.

3.3.3 The Surveys

An evaluative usability methodology was utilised as developed by Stevenson et al. [33] and further developed by the authors from an expert evaluation into an interview-based multiple choice user questionnaire with comment boxes to elicit quantitative and qualitative information. The questionnaire survey was conducted within 3 days in January 2014 that is 7–9 months after moving in—as the residents moved in two stages at the end of March and beginning of May 2013. All dwellings were visited using a 20 min survey with a 100% response rate. The usability survey grouped all user controls provided into 5 categories: MVHR, renewable controls, central heating and hot water, electricity and lighting, emergency and maintenance. It covered all environmental home controls—or 'touchpoints' of interaction between user and environmental technologies (Fig. 3.4). For each control same questions were asked to evaluate it against 8 criteria:

- clarity of purpose
- location
- understanding the operation
- ease of use
- labelling
- feedback
- fine-tuning
- understanding of the need for interaction

The survey questions gave three answer options: 'Yes', 'No', or 'I don't know'. The results of the survey, particularly when combined with onsite observations and conversations, gave the researchers a significant understanding of the home use learning stage, i.e. inhabitant's understanding and skills in interaction with controls as well as insight into quality of individual learning support provided, perceived technical skills and last but not least indicate issues with the design of the environmental controls. In the analysis of the quantitative results if the respondent answered 'I don't know' to questions: 'Is it clear what this control does?' or 'Is it obvious if you should interact with it?', this was regarded as indication of lack of understanding of a control and excluded that particular survey from further analysis of level of successful interaction. Importantly, information about prior experiences with similar systems and controls was obtained via an interview, and it would be beneficial to include it into the core usability survey in the future.

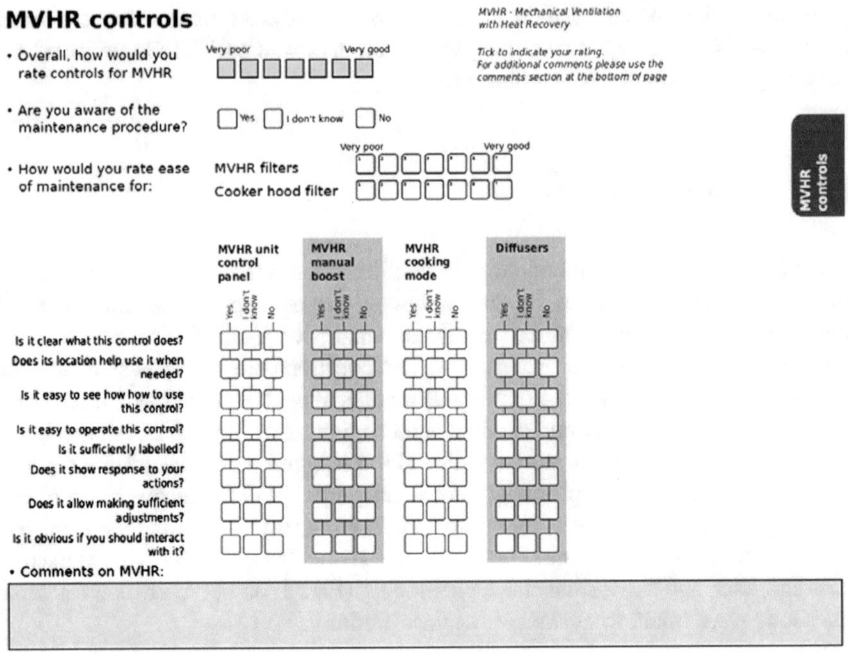
MVHR controls

MVHR - Mechanical Ventilation with Heat Recovery

• Overall, how would you rate controls for MVHR — Very poor … Very good

Tick to indicate your rating. For additional comments please use the comments section at the bottom of page

• Are you aware of the maintenance procedure? — Yes / I don't know / No

• How would you rate ease of maintenance for: — Very poor … Very good
MVHR filters
Cooker hood filter

MVHR controls

	MVHR unit control panel	MVHR manual boost	MVHR cooking mode	Diffusers
	Yes / I don't know / No	Yes / I don't know / No	Yes / I don't know / No	Yes / I don't know / No
Is it clear what this control does?				
Does its location help use it when needed?				
Is it easy to see how how to use this control?				
Is it easy to operate this control?				
Is it sufficiently labelled?				
Does it show response to your actions?				
Does it allow making sufficient adjustments?				
Is it obvious if you should interact with it?				

• Comments on MVHR:

Fig. 3.4 Usability survey MVHR page

The Building Use Survey (BUS) survey is a well-tested occupant feedback method which covers comfort and control as well as occupants wider needs [34, 35]. It was deployed on all 20 dwellings with a 100% participation rate.

The quantitative data from monitoring was managed using Thermodata software and Access. The data from surveys, meter readings was processed into Excel and together with monitoring data was included into database developed in Server Express to correlate results. Graphs to illustrate the results were produced in Excel. The interviews and other qualitative data were managed using NVivo 10 in order to code relevant categories of response.

3.4 Understanding the Key Home Use Learning Challenges

LILAC was set up around low impact and affordable living values. The community made decisions about building materials and technology selection guided by those crucial values. All homes are built with ModCell pre-fabricated timber and straw construction system which has received extensive government research funding in the UK [31]. The external walls and roofs use pre-fabricated straw timber panel and concrete ground floor slabs form a base. The mechanical and electrical (M&E) systems in each dwelling include the same MVHR—a heat recovery unit combined with a cooker-hood, gas central heating controlled by room thermostat, thermostatic radiator valves and photovoltaic panels. The flats have gas combi-boilers with 24 hours manual central heating programmers. The houses have a hot water heating system consisting of unvented hot water storage tank heated either by solar thermal panels or gas or an immersion heater as a backup. A heating/hot water programmer is also provided.

Interviews carried out with each household over a year after moving into the development, but before final BPE feedback was given, revealed that they expected the energy consumption to be low due to the special features of the building fabric and other technologies. This was verified with energy use lower compared to the previous accommodation that occupants were in. Some residents estimated a 90% reduction compared to their previous energy bills after just over six months of living in the development [36]. Once such energy savings have been assumed by inhabitants, efforts to further limit the energy use in low energy housing can be viewed by occupants as being less effective in terms of environmental benefit, according to the law of diminishing returns, than focusing on lowering impact from other areas like food or transport [37, 38]. This can, however, have other unforeseen consequences and was certainly the case at LILAC where relatively little attention at community level was initially given to supporting the process of home use learning of each household. However, satisfaction and capacity to interact with the MVHR system is seen as a critical challenge in air tight well insulated homes with low background noise levels where inhabitants easily notice and consider as disturbing noise or cold drafts and if unable to react differently tend to switch the system off which is against system's design intention [39–42]. The effects of this in relation to LILAC are presented in Sect. 3.8.

3.5 Analysing Home Use Expectations, Prior Experiences and Skills

Results from usability survey indicate varying levels of successful interaction for different controls. A mean number of 'Yes' answers for each control (maximum 8) was established from those surveys where respondent showed basic understanding of a control (Fig. 3.5). In the Usability survey the respondents were asked to indicate their perceived skills in using technical devices on a scale 1–7. Figure 3.5 also illustrates impact of this factor. Where occupants indicated a high level of skill this equated to a higher level of successful interaction with most controls, particularly those which are rarely used or are technically more advanced. The squares show mean 'Yes' results for the whole population whereas the green triangles show mean 'Yes' answers from those with positive self-evaluation of technical skills—scoring above 4 on 1–7 scale. Interestingly, the six lowest rankings of successful interaction are taken by MVHR, PV and solar thermal controls, i.e. for those systems which users had no prior experience of or were infrequently used. Top

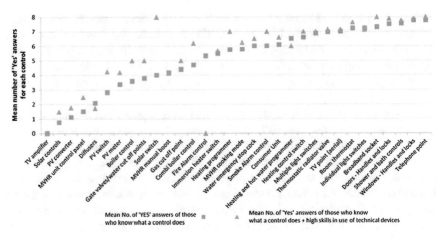

Fig. 3.5 Usability survey results: mean level of successful interaction with each control

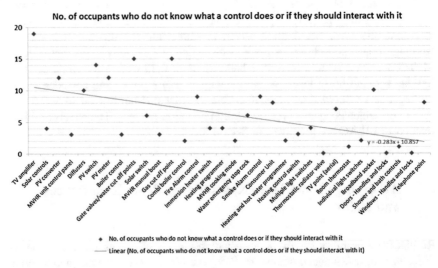

Fig. 3.6 Usability survey results: no. of occupants who do not know what a control does or if they should interact with it

scores are for controls well tested and understood in previous accommodation like door handles, windows and telephone points etc.

User understanding also varies for different controls as shown in Figs. 3.6 and 3.7 respectively showing number of households lacking understanding of a control (red diamond) and number of households with prior experience with a control. None of the case study residents had prior experience with MVHR, solar thermal or PVs.

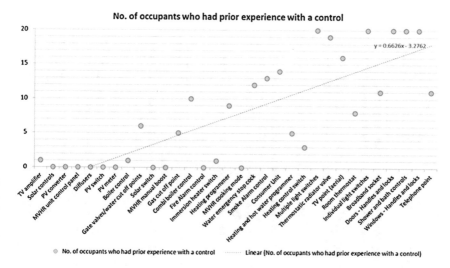

Fig. 3.7 Usability survey results: no. of occupants who had prior experience with a control

3.6 Provision of Individual Home Use Learning Support

LILAC did not commission a bespoke professional home handover procedure for its members, which is typically delivered by larger developers. Instead it invented a pioneering handover procedure to suit its particular circumstances. This involved the subcontractor responsible for M&E engineering and installation instructing selected LILAC members on the skills and understanding needed to make use of the systems installed as intended—to introduce users to the designer's model [24]. The plan was for these LILAC members to then pass on their knowledge to all other members. As a result cost savings would be made compared to a typical handover procedure while the skills and understanding of home use would be shared within the community. This home demonstration tour and subsequent dissemination was an explicit example of collective learning. The whole process was shadowed by the authors as a form of action research and resulting recommendations from the authors were fed back to LILAC, including a SWOT analysis of the process (Table 3.1). The main issues identified were lack of hands-on practice opportunity for participants (little reproduction), difficulty with providing a focused atmosphere —the excitement of the 'moving-in' day and relative group inattentiveness to the demonstration compared to 1-1 sessions, as well as lack of demonstrator M&E expertise leading to some misconceptions being passed on. Contradictory advice was given by the demonstrators on basic concepts of MVHR system operation, e.g. whether windows should be permanently closed or opened on hot days, or whether it was acceptable to interact with MVHR control panel to manually manipulate air flow levels in specific circumstances. The demonstrator was also unaware that the installed MVHR system was equipped with a humidistat which automatically

Table 3.1 SWOT analysis of LILAC innovative home handover process

	Favourable for LILAC handover procedure	Unfavourable for LILAC handover procedure
Internal	**Strengths** • Strong intentional community with set of values that each member signs up to • MEMBERS obliged to dedicate 2/4 h weekly for the community—can be spent on preparing and specialising in LILAC handover procedure • Additional time commitment by a member for LILAC's residents mutual benefit can be paid by the community • Diversified group of people with capacity to obtain skills and knowledge and reach innovative solutions • High communication skills within the group • Eagerness to learn, share knowledge and be as self-reliant as possible • Co-operation in terms of one maintenance contract to cover all the homes may lead to cost reduction	**Weaknesses** • Currently lacking M&E expertize within the group • Developing bad habits in terms of using homes as a result of lack of proper training • Lack of a good quality home user guide for reference • Lack of clear design intentions visible in terms of M&E user control locations • Limited resources for professional training for mutual benefit within the group • Limited time members can devote to additional tasks • Trust among the residents may weaken their critical view on information they are given. As there is no M&E expert, everything should be double checked • Lack of detailed repeatable procedure may result in variations in handover content
External	**Opportunities** • Avoiding the cost of paying an external handover/M&E specialist • All the M&E systems installed in LILAC have been installed somewhere else before—tracing precedent cases, particular in co-housing developments, and learning their lessons might be useful • Specialist trainings could be made available to develop M&E expertize within the group • Internet gives unprecedented opportunity for problems solving, gaining and storing information • Working with the Co-housing Network to develop bespoke solutions for Co-housing as a typology	**Threats** • Complexity of M&E systems installed could lead to failure and health risk due to lack of understanding • Degree of intended user's control over the systems not always clearly communicated to the user resulting in misuse • Industry trend to make the M&E systems user rely fully on specialists against LILAC aspirations of autonomy • Existing manufacturer's manuals too general or too specific for users (they are intended for specialists)—a learning gap for self-educating in M&E systems • Following unreliable advice and spreading wrong information to others result of lack of expertize in LILAC as a result of lack of expertize

adjusted air flow level according to the RH. All of these factors can lead to poor understanding and resultant 'bad habits' formation, unless further action is taken to resolve this type of misunderstanding.

The initial Home User Guidance (HUG) issued at an early occupancy stage contained only basic generic information about the M&E systems installed leaving occupants unable to retain the necessary information in order to properly learn how to use the systems. Installation manuals were also provided alongside the HUG but few occupants found them useful. As a result the HUG's guidance on operating the home and securing maintenance was negatively rated by all LILAC members. Out of 6 case study dwellings examined in detail in Sect. 3.8, the handover process covered only 3 dwellings meaning that half of them were inhabited without any induction process at all and had a poor HUG. The steep home use learning curve in early occupancy was thus mainly an individual learning process and resulted in varied ventilation strategies being adopted by different users, some working more successfully than others.

3.7 Decision-Making, Skills and Understanding Related to Home Use

In LILAC the key areas of concern that members actively embrace on a daily basis are expressed through the community's governance structure focused on a central hub which has oversight of 9 task teams:

- finance
- landscape
- maintenance
- food
- process
- membership
- learning
- community
- common house

Each member is encouraged to actively participate in a task team and each team has a spoke that reports back to the hub to ensure effective communication across the society. Individual time commitment is not logged and the system works based on mutual trust of equal effort but varied time resources. There is a non-hierarchical decision making process based on formal consensus decision-making at general meetings, based around for levels of decisions -routine, significant, major and emergency. Majority voting can be used as a fall-back position if the consensus decision-making process has not worked after three meetings. In the first two years of LILAC's life this has not occurred. Issues of ecological foot-printing captured the community's attention mostly at the early stage of group formation. It was then when LILAC identified a set of key factors: construction fabric, home energy and water use, food, transport and consumer goods [28]. These factors underpinned design brief and internal policies, e.g. 3 secure bike sheds were built and the community set a limit of maximum 10 cars for all the 20 households. Low energy and water use was to be achieved through highly insulated air-tight buildings supported by advanced mechanical and electrical (M&E) systems which were specified by a consultant initially and later designed and installed by subcontracted engineers. There was a general trust that these early decisions result in a low ecological footprint and there is no need for a dedicated Low Impact Living task team to further improve it. Until the BPE process feedback was disseminated by the authors, there was no awareness of over 5-fold difference in daily water consumption per person within the development (min is 42 l/per person.day and maximum is 223 l/per person.day) or 4-fold variation in gas and electricity consumption between households. According to the results of the usability survey the understanding and skills to interact with the controls also varied significantly across the 20 LILAC dwellings (Fig. 3.8). This indicates high potential benefit of further skills transfer within the development through additional social learning and training, and the formation of a dedicated on-site energy task team. Learning opportunities offered by the BPE via occupants meeting with the authors and learning from each other via the authors have been picked up by the community due to its responsive decision making process. As a result of the study, maintenance task team remits have been extended at an annual community meeting to encompass responsibility of the team to ensure that everyone has an understanding of how the houses function. Through the BPE, members have also become more aware of their surroundings, scrutinising more closely elements such as the airtightness of the building fabric and the functionality of mechanical and electrical controls. In the occupancy phase LILAC has also upgraded various heating controls, as an integrated programmable thermostats were not provided. An overall finding from the LILAC project, is that the attention of residents was too focused on the low impact potential of the superstructure and building fabric to the detriment of the wider mechanical and electrical installations which would be built around it. This would be usefully remedied through embedding an expert integrated energy engineer during the development and early occupation phase.

3.8 Correlation of Results for MVHR in Relation to Clarity of Use

A detailed analysis correlating results for internal moisture levels achieved and satisfaction and capacity to control mechanical ventilation with heat recovery (MVHR) in six selected dwellings demonstrates the need for home use improvement and the capacity of collective learning to deliver the change. The selection covers the typological and demographic variation in the development and includes

Understanding of controls:

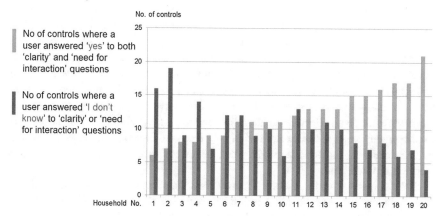

Fig. 3.8 Variation between households in number controls that are understood (*green*) and those that are not (*red*)

3 flats and 3 houses inhabited by either two adults or 2 + 2 families. The selection keeps the fabric variables to a minimum by a selection of ground floor 1–2 bedroom flats and 3–4 bedroom semi-detached houses.

Table 3.2 compares results from the usability survey with the BUS survey and RH monitoring for all the case study dwellings to help analyse the need and capacity for collective learning related to efficient home use. Results from the usability survey reveal that there is a wide variation in skills and understanding of control use compared to perceived degree of control expressed in the BUS survey (Table 3.2) which tends to undermine the BUS results. The usability survey shows that some occupants, who have never interacted with most of controls, indicate no problems and perceive the systems to be 'working ok'. Others, who do attempt to adjust system to their needs (e.g. reducing MVHR night time low air flow to save energy and reduce noise), are more critical of missing features and desire more control (Table 3.2). On-site observation of home use patterns and informal conversations identified that those who did not feel competent to use the controls provided, and at the same time were not satisfied with the system functioning, found their own ways to cope with the situation, which did not necessarily correspond with design intentions. In case study No. 1 the residents mentioned a persistent cold draft from the MVHR supply diffusers in bedrooms. As they couldn't control this they turned the MVHR off and opened the windows during the day instead, although security reasons at ground floor flat restricted the time when windows were actually open. This home use pattern extended beyond free-running mode towards November. Windows opening behaviour was not directly monitored in this study but rather based on an extended BUS survey and on-site observations every 8–9 weeks within 3–4 subsequent days between June 2013 and July 2014. Custom questions developed by the authors and added to BUS survey captured the relationship between heating, ventilation and windows opening behaviour in 4 seasons. In case study No. 2 the residents turned the MVHR off at night because they found the noise too disturbing. They did not try to set MVHR on lower setting to reduce the noise as residents did successfully in dwelling No. 4 and No. 5. The information about the lower MVHR setting option was eventually shared by a LILAC member, who figured out that possibility, with all the members via internal email—another example of a collective learning attempt. However it was picked up and applied by only one other member—someone who had also declared very high skills for using technical devices in the usability survey. This finding suggests that email may not be an effective means of collective learning about technical home use improvement issues for those not technically inclined. Physical demonstrations with guided hands-on experience might work better for them as a more encouraging learning environment.

The comparison of the usability survey results and on site observations with RH monitoring results reveals another hidden issue. In the two dwellings with low occupant skills and prolonged periods of the MVHR being off, the monthly average RH levels monitored in bedrooms and bathrooms exceed the 65% RH threshold in the heating season set to prevent mould growth (see Table A2 in [42]). The average RH in No. 1 for November 2013 was 72 and 75% in No. 2, whereas it was within the 55–60% range in all the dwellings where occupants had either a high under-standing of the MVHR system or a low understanding but satisfaction with its performance (Table 3.2). As the homes were not initially intended to be mechan-ically ventilated excellent natural cross-ventilation opportunities exist in the kitchen/living area. However, the bathroom window is not easily accessible, thus switching the MVHR off increases the risk of high moisture in the bathroom. This suggests, critically, that a lack of skills and understanding to control these systems may lead to users developing home use patterns which lower the quality of internal environment. This finding requires further investigation as the MVHR air flow rate check was performed in only 2 of the 6 case study dwellings—the ones performing well in terms of RH. Major issues in terms of design and installation identified for all the dwellings, however, do not explain all the variation captured through RH monitoring. It may be possible that poor performance is due to a system fault. However, it is clear that the occupants are neither aware of high humidity nor of the potential for improvement of their control of internal environment, which is also a contributing factor. The BUS survey shows that the users in the two humid flats rate air humidity in winter in their dwellings as dry—'2' on a 1 (dry)—7 (humid) scale (Table 3.2). The survey also shows that those who have the most excessively high humidity levels paradoxically say they have full control of ventilation system (Table 3.2). These findings suggest that some LILAC members are in a state of 'unknown unknowns' that are difficult to reveal without them confronting and comparing their own way of coping with home use issues with those of others.

As this study suggests the increased internal moisture levels in households turning the MVHR off and opening windows instead, is linked with the seasons and variations in the outside RH levels. The problem does not seem to exist in the hot periods when many of the analysed households find opening the windows and turning the system off the only effective way to cope with the heat. Moreover

Table 3.2 Comparison of results for case study dwellings

Case study dwelling	No. 1	No. 2	No. 3	No. 4	No. 5	No. 6
Average RH November	72% > max. 65%	75% > max. 65%	60%	60%	56%	55%
BUS survey rating						
Winter comfort 1 (dry)– (humid)	2	3	2	2	4	3
Personal control over ventilation: 1(no control)–7(full control)	7	6	6	4	6	5
Key comments		MVHR often off at night because of the noise. We often open windows and have MVHR off to save energy		Don't like lack of control of MVHR. No instructions given for MVHR. Have read installation manuals to understand better. MVHR on a timer [low flow] to reduce noise in the night	Would like to get smart app for more control	Comfy temperatures and humidity MVHR is on all the time except when it's hot and we open the windows
Usability survey						
Skills and understanding of controls' use	Low	Low	Medium	High	High	Low
Key comments	Our controls were set up at the beginning and never were set up again because they've been working ok		MVHR unit control panel menu options seems limited + are not very clearly understandable for a non-technical home user	All are standard controls I have used before. CH/hot water programmers lack an extend feature which is very useful when you want to put it on for half an hour		
On site observations and conversations						
Issues with MVHR performance	Cold draft from bedroom supply point	Noise				
		–	–	–	–	–

turning the system off evidently lowers energy consumption throughout the free-running season which is regarded as positive by the environmentally conscious inhabitants. Curiously, the MVHR manufacturer advises, however, that the system is designed for 24/7 use and should never be turned off [43]. Judgement on how to respond to such dilemmas in a specific context is a part of home use learning process. Comparing varied home use practices deriving from different judgements within a single development has the invaluable potential to identify 'best practice' and unlock the home use improvement potential. This seems to be one of crucial benefits of expanding the BPE research with social learning framework elements to enhance collective learning. The collective learning can help to link the designer and user models [19] and may be useful in developing co-evolutionary adaptivity [44] if circumstances not covered by the design model are experienced.

3.9   Discussion of Barriers and Opportunities for Collective Learning

This study shows that LILAC constitutes a very well integrated community with a high level of communication skills. Out of 20 BUS questionnaires returned 17 mention highly positive experiences of leisure, contact or support that the community offers. Members have numerous occasions to meet up: in the Common House, in the common open spaces between the blocks, and in the allotments. All members have access to the community notice board next to their communal post area as well as internal mail and an on-line forum. The main means of information dissemination on home use to all members is via email. However, as discussed earlier, this means of communication may not be an efficient way to introduce inhabitants to home use adjustments. LILAC members were actively involved in developing an energy strategy to achieve the 'low impact living' goal and made recommendations to the contractor. However, since moving in the only collective discussions focused exclusively on home use issues have been the home demonstration tour and the feedback meetings on the POE findings organized by the authors.

Interestingly, a wide range of variety in understanding of the homes' performance can be observed across the case study households even where they have an active member in the hub group (No. 5 and 6) or maintenance team (No. 3 and 5). This is because a single task team has members with both high and low levels of understanding. This suggests that collective learning of home use within these environments has not had the opportunity to be fully deployed due to lack of capacity and despite the need for it to happen.

Table 3.3 summarises the barriers and opportunities for collective learning identified through the initial SL analysis of LILAC. It ties in with the SL conceptual framework by referring to selected elements of social learning process (Fig. 3.3):

Table 3.3 Collective learning barriers and opportunities as observed in LILAC

Category	Collective occupant learning identified barrier	Identified opportunity/potential
Observational learning	Attention: no clearly defined specific efficient home use goals + broad scope of outreach activities—no focus on home use improvement Poor HUG—individual learning more difficult	Group meter readings and dissemination potential due to trust —'best practice' approach possible within the group High skills of using controls within the group + intention to develop dissemination procedures
Motivation	Limited link between low energy/water consumption and low bills: energy tariffs with high standing charges and low unit cost + metered water consumption leading to higher bills than before even if lower use Not all users involved benefit equally from engaging in collective home use learning process	Varied level of understanding the bills—potential for greater knowledge dissemination Strong pioneering ethos and 'best practice' ambition Intentional community around values of low impact living, self-reliance and connectivity—connect to SL
Information flow/channels	Information overload leading to limited interest or missing out + limited time resources No home use focused collective learning events organized Social pressure on energy improvement limited to respect privacy of each household	Varied learning environments available: **Physical**: interaction intended by the community and enhanced by design: site layout and provided facilities (e.g. common house) **Internet based**: internal email and on-line forum, maintenance bespoke software managed by maintenance task team with on-line access to all members
Identifying areas for development	Poorly communicated designer model + unclear energy goals—lower individual awareness of the need/capacity for home use improvement Maintenance task team oriented towards trouble shooting + securing basic maintenance—lack of forward thinking 'home use improvement' agenda Low expectations due to previous experiences	**Performance**: varied energy efficiency and achieved internal environment (e.g. RH) between dwellings **Skills**: complex M&E systems installed—provide experience with the provided technology and support for individual home use learning **Varied** actual **user models** of coping with technology—potential for revealing 'unknown unknowns'
Seeking solutions	No M&E specialist within the group + limited resources for external training	High problem solving capacity within the group

- motivation
- observational learning
- information flow
- identified areas for development
- seeking solutions.

The main issue appears to be lack of early community focus on home use improvement and the time to engage with this. During the early phase of occupation, residents of LILAC were mainly focused on resolving project defects, settling into their new lives and finalising policies and procedures to run the self managed cooperative society. As explained earlier, motivation is directly linked with an awareness of goals (e.g. relevant benchmarks), understanding the saving potential on water and energy bills and being able to follow one's own consumption. A basic difficulty is the pioneering character of this housing development, which can tend to privilege outward facing communication at the expense of certain areas of internal communication related to improving home use, particularly where there is a collective case of 'unknown unknowns' related to new technology. However one opportunity might be for national networks supporting cooperative and cohousing projects to facilitate collective learning based on best practice national or international precedents within the group which can allow LILAC to remain pioneering and outward facing while at the same time addressing internal concerns. Surprisingly, the potential for lowering energy and water bills any further is questioned in some of the BUS survey comments. Even though 85% of LILAC households claimed a lower consumption of energy compared to living in their previous accommodation, only a few commented on the high standing charge that negated the expected financial benefit. High standing charges undermine the effect on energy bills resulting from improved home use patterns. The self-awareness of household energy and water consumption varied significantly across the community. Ease of access to meters varied: water meters were the most difficult to read for all users and some claimed they had not received their water bill yet.

In a large UK survey Seyfang et al. [45] identify a number of different methods used by community groups to improve their understanding of energy use, demonstrating that a greater emphasis is put on engagement and dissemination rather than learning per se. They note the quality of the group concerned as decisive in terms of its success in achieving its aims, followed by the skills available within the group. A critical weakness identified was the lack of time available to the group. The results for LILAC tie in with Seyfang's findings, which show that communities tend to be more outward facing than inward when trying to learn about energy use. It is clear from LILAC's task team structure that understanding home energy use is only one of a number of key concerns related to low impact living and is not a top priority. With limited time, members have to balance their collective desire to understand the various aspects involved in developing low impact living with their need to communicate to others and their need to try and understand how their own homes perform. A dedicated 'maintenance task team' manages incident reports for all the households organizing necessary repairs with the help of bespoke software accessible to all members. The system takes care of all trouble shooting needs but little time is left for forward planning by reflecting on current home use patterns and possibilities to facilitate further improvements. Some team members seem to have acquired enough expertise and experience to know the limitations of the systems installed and have control over their internal environment but this know-how is not yet widely disseminated.

As this study suggests individualised learning processes lead to varied results: some households find on their own the optimum way of using their homes while others end up in a state of 'unknown unknowns' with engrained misconceptions and a lack of awareness that improvements are readily available.

Significant differences established in internal home environments can be linked with varied levels of understanding and skills to control ventilation systems in the case study development. This suggests a high home use knowledge sharing potential within the development that is not yet exploited.

Some households, unaware of their lack of knowledge, do not look for solutions or improvements in their use of technology. Collective home use learning through sharing experience and being exposed to different ways of using the technology provided within newly inhabited developments is argued to have the potential to challenge such an issue.

As this study shows this has led to substantial differences in understanding and skills to interact with mechanical and electrical systems provided and varied home use practices.

3.10 Conclusions

This paper has demonstrated an initial conceptual framework for collective learning of home use. The framework sets out the scope of analysis in a case study housing development to illustrate the home use learning challenges and identify barriers and opportunities of collective learning process in a specific context. A case study of an innovative low carbon housing development in England indicates strong potential in the community for collective home use learning which has not yet been fully developed.

A major opportunity of learning through social interaction within a community is to unlock individual misperceptions of control through exposure to different ways of approaching and dealing with the same comfort/energy/water saving issues by other people. This can lead to discovering a variety of system misuses or failures that are often just 'coped' with—with occupants unaware that some technical features may be designed to work differently from the misaligned habits they have adopted. Without such home use experience exchange there is a long term risk of occupants being unaware that something is wrong or a lack of understanding resulting in the system being used incorrectly—this can lead to the development of long term bad habits based on wrong assumptions. This study shows a general lack of awareness of the differences in home use skills and understanding between households in a particular housing development. In the UK, there is a culture of people not always being willing to show each other that they don't understand how to use their home—a certain 'house proud' factor that can lead to numerous occasions for knowledge and skills exchange never being used to improve home use understanding. This suggests that the home use collective learning needs careful facilitation even in a well-integrated community, as evidenced by the initial inefficiency of internal dissemination of know-how in the LILAC development. Building Performance Evaluation (BPE) feedback given to the residents has become such facilitating factor.

Interestingly, the level of occupant control over their internal environment is not clearly linked with the household's total energy consumption in this particular case study. The most highly motivated households focused on energy and water saving are not just those with highest skills in controlling the technology. When there is a lack of understanding of what to expect from a new system, (such as the analysed MVHR, and how to control it) combined with a lack of satisfaction with its performance, it is not necessarily the level of consumed energy that increases but the quality of internal environment that decreases. More importantly this decrease stays unnoticed presenting potential health risk to inhabitants.

At an individual level, the benefits of getting involved in home use learning may not be obvious—unless there are major comfort issues experienced or high energy bills—however, neither is the case in LILAC and so an additional incentive may be required. One way to increase individual engagement is through focused community wide engagement. A barrier to achieve that is this not all occupants are involved equally or directly benefit from engaging in collective home use learning process covering the whole community. The most advanced users gain least but without them the social learning process cannot not work effectively. It is recognised in management theory that in the complex and dynamic world we live in today structuring the 'unknown' in order to be able to take action, referred to as 'sensemaking', is a genuinely collective process. It requires co-operation of a group of individuals with varied viewpoints and skills [46]. The home environment has recently become part of a profound and dynamic transition towards zero-energy buildings and is also prone to the difficult to predict effects of climate change. 'Co-evolutionary adaptivity' defined as developing buildings' and users' ability to respond to unknown changes through time [44] bears resemblance to 'sensemaking' which suggests its collective nature. In this case study, technology related learning was collective at points but did not cover all members. Instead exclusive 'communities of practice' [47, 48] emerged, consisting of the most technically keen inhabitants, where the home use learning progressed. Wider dissemination of results of that learning occurred as a direct result of the BPE process and findings.

Acknowledgements The authors gratefully acknowledge the funding provided for research informing this study by the EU FP7 Marie Curie Intra European Fellowship programme through the BuPESA project, as well as the generous time given by the occupants involved in these case studies.

References

1. Mohamed, A., Hasan, A., Siren, K.: Fulfillment of net-zero energy building (NZEB) with four metrics in a single family house with different heating alternatives. Appl. Energy **114**, 385–399 (2014)
2. Stefanovic, A., Bojic, M., Gordic, D.: Achieving net zero energy cost house from old thermally non-insulated house using photovoltaic panels. Energy Build. **76**, 57–63 (2014)
3. Szejnwald Brown, H., Vergragt, P.: Bounded socio-technical experiments as agents of systemic change: The case of a zero-energy residential building. Technol. Forecast. Soc. Change. **75**, 107–130 (2008)
4. Yua, Z., Fungb, B., Haghighata, F., Yoshinoc, H., Morofsky, E.: A systematic procedure to study the influence of occupant behavior on building energy consumption. Energy Build. **43** (6), 1409–1417 (2011)
5. Gill, Z., Tierney, M., Pegg, I., Allan, N.: Low-energy dwellings: The contribution of behaviours to actual performance. Build. Res. Inf. **38**(5), 491–508 (2010)
6. Gram-Hanssen, K.: Efficient technologies or user behaviour, which is the more important when reducing households' energy consumption? Energ. Effi. **6**, 447–457 (2013)
7. Janda, K.: Buildings don't use energy: People do. Architectural Sci. Rev. **54**, 15–22 (2011)
8. Her Majesty's Governement: National Strategy for Climate and Energy. The Stationary Office, Norwich (2009)
9. Stern, P.: Individual and household interaction with energy systems: Toward integrated understanding. Energy Res. Soc. Sci. **1**, 41–48 (2014)
10. Gram-Hanssen, K.: Residential heat comfort practices. Build. Res. Inf. **38**(2), 175–186 (2010)
11. Gill, Z., Tierney, M., Pegg, I., Allan, N.: Measured energy and water performance of an aspiring low energy/carbon affordable housing site in the UK. Ener. Build. **43**, 117–125 (2011)
12. Rohracher, H., Ornetzeder, M.: Green buildings in context: Improving Social Learning processes between users and producers. Built Environ. **28**(1), 73–85 (2002)
13. Day, J., Gunderson, D.: Understanding high performance buildings: The link between occupant knowledge of passive design systems, corresponding behaviours, occupant comfort and environmental satistafaction. Build. Environ. **84**, 114–124 (2014)
14. Ajzen, I.: The theory of planned behaviour. In: Van Lange, P., Kruglanski, A., Higgins, E. (eds.) The Handbook of Theories of Social Psychology, vol. 1, pp. 438–460. Sage, London (2011)
15. Chatterton, P.: Towards an agenda for post-carbon cities: Lessons from LILAC, the UK's first ecological, affordable, cohousing community. Int. J. Urban Reg. Res. ISSN 0309-1317 (2013)
16. Leaman, A., Stevenson, F., Bordass, B.: Building evaluation. Build. Res. Inf. **38**(5), 564–577 (2010)
17. Mallory Hill, S., Preiser, W.F.E., Watson, C.: Enhancing Building Performance. Wiley Blackwell, Oxford (2012)
18. Way, M., Bordass, B.: Making feedback and post-occupancy evaluation routine 2: Soft Landings—involving design and building teams in improving performance. Build. Res. Inf. **33**(4), 353–360 (2005)
19. Glad, W.: Housing renovation and energy systems. Build. Res. Inf. **40**(3), 274–289 (2012)
20. Berry, S.: Inspiring low-energy retrofits: The influence of 'open home' events. Build. Res. Inf. **42**(4), 422–433 (2014)
21. Gupta, R., Barnfield, L., Hipwood, T.: Impacts of community-led energy retrofitting of owner-occupied dwellings. Build. Res. Inf. **42**(4), 446–461 (2014)
22. Gertler, M.: Technology, culture and social learning: Regional and national institutions of governance. In: Gertler, M., Wolfe, D. (eds.) Innovation and Social Learning, pp. 111–135. Palgrave Macmillan, New York (2002)
23. Carmona-Andreu, I., Stevenson, F., Hancock, M.: Low carbon housing: Understanding occupant guidance and training. In: Hakansson, A., Hojer M., Howlett, R., Jain, L. (eds.)

Smart Innovation, Systems and Technologies. Sustainability in Energy and Buildings. Proceedings of 4th International Conference on Sustainability in Energy and Buildings. pp. 545–554. Springer, Heidelberg (2012)

24. Norman, D.: The Design of Everyday Things. MIT Press, London (1998)
25. Badura, A.: Social Learning Theory. Prentice-Hall, New York (1977)
26. Sørensen, K.H.: Learning Technology, Constructing Culture: Socio-technical Change as Social Learning. STS Working Paper No. 18/96, Centre for Technology and Society, University of Trondheim, Trondheim (1996)
27. Mallory-Hill et al. (eds.).: Enhancing Building Performance. Wiley-Blackwell, Chichester (2012)
28. Biggs, R., Schlüter, M., Biggs, D., Bohensky, E.L., Burn Silver, S., Cundill, G. et al.: Toward principles for enhancing the resilience of ecosystem services. Annu. Rev. Environ. Resour. 37, 421–448 (2012)
29. Yin, R.K.: Case study Research: Design and Method, 2nd edn. Sage, London (2009)
30. Chatteton, P.: Low Impact Living—A Field Guide to Ecological Affordable Community Building. Earthscan, Oxon (2015)
31. Wall, K., Walker, P., Gross, C., White, C., Mander, T.: Development and testing of a prototype straw bale house. Proc. Inst. Civil Eng. Constr. Mater. 165(6), 377–384 (2012)
32. Baborska-Narozny, M., Stevenson, F.: Performance evaluation of residential architecture. In: Chartonowicz, J. (ed.) Advances in human Factors and Sustainable Infrastructure. Proceedings of the 5th International Conference on Applied Human Factors and Ergonomics. pp. 109–118. AHFE, Krakow (2014)
33. Stevenson, F., Carmona-Andreu, I., Hancock, M.: The usability of control interfaces in low-carbon housing. Architectural Sci. Rev. 56(1), 70–82 (2013)
34. Cohen, R., Standeven, M., Bordass, B., Leaman, A.: Assessing building performance in use 1. Build. Res. Inf. 29(2), 85–102 (2001)
35. Chiu, L.F., Lowe, R., Raslan, R., Altamirano-Medina, H., Wingfield, J.: A socio-technical approach to post-occupancy evaluation: Interactive adaptability in domestic retrofit. Build. Res. Inf. 42(5), 574–590 (2014)
36. Chatterton, P.: Interviewed by Modcell. http://www.modcell.com/news/modcell-cuts-energy-bills-90/
37. BioRegional.: What makes an eco-town? CABE (2008). http://www.bioregional.com
38. Galvin, R.: Why German homeowners are reluctant to retrofit. Build. Res. Inf. 42(4), 398–408 (2014)
39. Balvers, B., Bogers, R., Jongeneel, R., Kamp, I.: Mechanical ventilation in recently built Dutch homes: Technical shortcomings, possibilities for improvement, perceived indoor environment and health effects. Architectural Sci. Rev. 55(1), 4–14 (2012)
40. Stevenson, F., Fewson, K., Johnson, D., Yeats, A.: Lancaster Co-housing project part 8: Post-occupation building performance evaluation. Green Build. 23(1), 24–35 (2013)
41. Brown, C., Gorgolewski, M.: Understanding the role of inhabitants in innovative mechanical ventilation strategies. Build. Res. Inf. 43(2), 210–221 (2015)
42. UK Building Regulations: Approved Document Part F, Ventilation. Crown Copyright, London (2010)
43. Vent-Axia.: Heat recovery ventialtion—home owners ventilation system guide. http://www.vent-axia.com/files/pdf- downloads/Sentinel_Kinetic_User_Guide_402270.pdf
44. Stevenson, F., Baborska-Narozny, M.: Designing resilient housing for co-evolutionary adaptivity. In: Nicol, F., Roaf, S., Brotas, L., Humphreys, M. (eds.) Proceedings of 8th Windsor Conference: Counting the Cost of Comfort in a Changing World, pp. 436–445. NCEUB, Windsor (2014)
45. Seyfang, G., Park, J., Smith, A.: A thousand flowers blooming? An examination of community energy in the UK. Energy Policy 61, 977–989 (2013)
46. Ancona, D.: Sensemaking: Framing and acting in the unknown. In: Snook, S., Nohira, N., Khurana, N. (eds.) The Handbok for Teaching Leadership, pp. 3–21. Sage, London (2012)

47. Wenger, E.: Communities of Practice: Learning, Meaning, and Identity. University Press Cambridge, Cambridge (1998)
48. Gann, D.: Trading places-sharing knowledge about environmental building techniques. In: Cole, R., Lorch, R. (eds.) Buildings, Culture and Environment, pp. 37–57. Blackwell Publishing, Oxford (2003)

Chapter 4
An Archetype Based Building Stock Aggregation Methodology Using a Remote Survey Technique

James Pittam, Paul D. O'Sullivan and Garrett O'Sullivan

Abstract Developing representative archetypes using a bottom up approach for stock modelling is an excellent tool for evaluating the overall performance of the building stock; however it requires detailed analysis of the various building types. Currently there is no detailed housing data base for Local Authority housing in Ireland that catalogues the housing stock according to geometric configuration and thermal characteristics for each typology. The aim of this chapter is to present a methodology to catalogue LAH stock and build a detailed housing stock data base. The GIS web based mapping application Google Street View is used to identify 18 house typologies across 36 LAH developments for Cork City in the South of Ireland, used as a dataset for demonstration of the methodology; a total of 10,318 housing units are counted and information subdivided into end of terrace, mid terrace, semi-detached, terrace lengths, orientation and elevation. This database then provides the base line assessment for building a stock aggregation model; the stock aggregation approach is used as a method to evaluate the energy performance of the building stock, beginning with analysis of individual house types; referred to as a 'bottom up approach'. Four representative archetypes are produced from the study and subsequently modelled using DEAP simulation software. 10,318 homes were identified, catalogued and statistically analysed within a number of weeks for the city council. This can be extended nationally very effectively with data bases now being constructed remotely rather than the challenges of physical mapping and surveying. Using a matrix based linear system for weighting parameters allows a large number of computationally efficient transformations of house types and parameters into archetypes. There is also flexibility in determining the grouping algorithm depending on the nature of the LAH studied. Investment in retrofit is highly justified in this area with large potential for reducing CO_2 emissions, the number of fuel poverty sufferers and victims of seasonal mortality due to thermally inefficient homes. The study suggests the method applied has scalable potential and is modular in structure facilitating wider adaptation.

J. Pittam (✉) · P.D. O'Sullivan · G. O'Sullivan
Cork Institute of Technology, Architecture, Cork, Ireland
e-mail: dr.jamespittam@gmail.com

© Springer International Publishing AG 2017
J. Littlewood et al. (eds.), *Smart Energy Control Systems for Sustainable Buildings*,
Smart Innovation, Systems and Technologies 67,
DOI 10.1007/978-3-319-52076-6_4

Keywords Local authority housing · Stock aggregation · Retro fit · Archetype · Fuel poverty

Nomenclature

Symbols

A	Area
V	Volume
S/V	Surface/volume ratio
h	Characteristic height
w	Characteristic width
d	Characteristic depth
S	Housing stock
a	Column vector
P	Parameter

Subscripts

j	Type
t	Length

Abbreviations

GE	Google Earth
GSV	Google Street View
LAH	Local authority housing
MT	Mid terrace
ET	End of terrace
SD	Semi-detached
RCM	Remote cataloguing method
RMM	Remote measurement method
BER	Building energy rating
DEAP	Dwelling energy assessment procedure
EM	Engineering method
SM	Statistical method

4.1 Introduction and Background

The residential sector is responsible for consuming 30% of global primary energy [1]. In Ireland this sector accounted for 27% of all primary energy used in 2011 [2]. Irelands housing stock is amongst the poorest performing in Northern Europe [3, 4]. Therefore tackling energy efficiency measures in the domestic sector is of extreme priority. A study carried out to examine CO_2 emissions found that the average Irish dwelling emitted 47% more CO_2 than the average UK dwelling and are 92% higher than the average EU-15 dwelling [5]. Ireland and the UK have the highest rates of seasonal mortality in Northern Europe despite having such mild winters; a number of studies have shown that this rate of mortality is directly linked to the thermally inefficient housing stock in Ireland and the UK [6]. Homes built between 1941 and 1979 have the highest absolute number of fuel poverty sufferers with 111,000 homes affected [7]. This associated age band is representative of the LAH stock presented in this chapter. In a report carried out by the SEAI it was found that those experiencing high levels of fuel poverty are more likely to live in semi-detached or terraced housing and is higher amongst tenants than owner-occupiers [8]. Plans to meet Kyoto targets for GHG reductions have obligated a number of countries to create detailed data bases of their building stock, and to model total energy use of representative buildings [9]. Improving energy consumption of existing buildings is an extremely effective method to reducing CO_2 emissions. Assessing the baseline performance of existing stock is a crucially important measure to designing retrofit solutions. The energy consumption in the residential sector is highly complex as there are a huge amount of variables and thermal characteristics which effect building performance. Understanding these effects is paramount to designing effective solutions. Analysis can be frustrated by lack of empirical data or by data which does not describe accurately the stock being represented. It was found that a pre-existing aggregated building database containing classification of LAH stock, geometric configuration and thermal characteristics was not in existence in Ireland's local authority housing departments.

The aim of this study is to use stock aggregation theory to build a detailed housing stock database for one Irish city council as a pilot study to test an archetype development methodology based on using a web based geospatial information mapping tool, and to build detailed representational archetypal buildings that reflect the entire stock under analysis. There are a total of 10,318 homes included in the study for the chosen council, Cork City. The aim is to use the archetypes to produce a baseline assessment of energy performance and to better understand the situation regarding Ireland's thermally inefficient housing stock. This will then aid in testing various permutations of retrofit solutions to help reduce fuel poverty and seasonal mortality in Ireland.

4.2 Overview of Existing Studies Using Stock Modelling Methodologies

A number of authors have presented methodologies which aggregate building stocks through the use of archetypal buildings to measure energy and resource quantities from regional to global levels. Stock modelling is used to evaluate total primary energy use and energy related impacts on a housing stock. It has been observed through literature that the choice of variables to describe energy use in a housing stock used to build archetypes seems to be based on author's discretion. A study carried out at an urban scale by Lechtenbömer [10], used two archetypes to describe the entire EU housing stock using construction periods, typologies and building size, but admitted significant uncertainties from lack of accurate statistical data. To understand the impact from the geometrical configuration of archetypes.

Lechtenbömer's method would lack that detail. Johnston [11] used 2 archetypes and a disaggregated bottom-up energy and CO_2 emission model of the UK housing stock and achieved 60% CO_2 emission reductions by the middle of this century using a weighted average dwelling approach to extract energy analysis. Johnston found that thermal characteristics of building fabric and efficiencies of heating equipment had a higher impact on energy use than dwelling type; however dwelling type in the context of this study is weighted heavier due to the impact of built form and cold bridging construction details on energy use and comfort. Knowledge of the composition of geometrical configuration is paramount to designing effective ret-rofit solutions. In contrast to Lechtenbömer, Firth and Lomas used 47 archetypes to describe the English residential housing stock using built form characteristics which are a key factor in annual space heating demand; and also using a weighted average dwelling approach for energy analysis [12]. A residential energy use model for Osaka city was developed by Shimoda using 43 variables made up through occupancy patterns, detached house types and apartment types [13]. The proposed method in this study is similar to a study carried out by Shorrock et al., who used survey data including U-values, physical elements of building and heating system equipment and efficiencies to develop 1000 archetypes to describe energy use of the UK housing stock [14]. We used a composite of geometrical configurations using a remote measurement method (RMM) to extract data and a BER database of 1551 homes to determine thermal characteristics. Similarly Parekh describes a method to develop archetype libraries through geometric configuration, thermal characteristics and operational parameters for energy use evaluation of the Canadian housing stock [15]. A study carried out by Famuyibo et al., applying a bottom-up method, describes 65% of the Irish housing stock by 13 representative archetypes using the Energy Performance Survey of Irish Housing (EPSIH) in combination with a comprehensive literature review to identify key deterministic variables which impact on energy use. The EPSIH database data was collected from a sample of 150 houses in 2005 which is slightly out of date and less representative of the housing stock when compared to using current surveyed BER results modeled using DEAP which take into account all recent upgrade works. Petersdorf uses five building types with 8 insulation standards and 3 climate zones to measure impacts on heating demand and CO_2 emissions [16], generating a total of 210 building models. The study found that the residential sector accounts for 77% of heating related CO_2 emissions of the EU housing stock with 23% being represented by non-residential buildings, with single family homes representing the largest group responsible for 60% of total CO_2 emission. The TABULA project used actual building types [9], however the use of representative building typologies in TABULA as opposed to using archetypes makes the extrapolation of typologies to arrive at stock aggregation not overly representative of stock. Swan and Urgursal developed an energy and emission model of the Canadian housing stock using a hybrid model consisting of a top-down and bottom up approach [17]; Annex 31 describes the combination of stock aggregation with regression models as an improvement in the forecasting ability of the model. The hybrid model can also be used to calibrate bottom-up models with top-down models [18].

In summary the uses of average dwelling types as opposed to models are used by a number of authors. The model method using a bottom-up approach is more appropriate for this study where a large concentration is on the impact from built form characteristics on energy use. Energy modelling the archetype gives a detailed representative house capable of describing losses through building fabric and helps pin point problematic areas. Data for construction details is also absent in a number of the studies due to the use of existing aggregated statistical databases. One of the strengths to the proposed method for aggregating archetype thermal characteristics using BER data, is that the ratings are based on detailed survey information and take into account any upgrade works to fabric or heating system as opposed to using historical aggregated statistical data, which is often out of date and lacking in detail.

4.2.1 Various Modelling Techniques

This section describes various stock modelling techniques with two main approaches identified namely top-down and bottom-up modelling [19]. Stock aggregation (bottom-up) and the top-down method are significantly different approaches to obtaining information on physical resource flows through a building stock; the choice of approach depends on a number of factors. Analysts also sometimes rely on a hybrid model which employs both methods [18]. The top-down approach regards the residential sector as an energy sink, and uses historical aggregated energy values; it applies regression on the energy consumption of the housing stock being studied and associated energy drivers. The strengths of the top-down model exist in its sole reliance on statistically historical aggregated data for housing records. This enables modelling of the entire residential sector to describe energy use by variables; which highlights potential upgrade areas to aid large scale optimum decision making. It is a faster and more affordable method because of reduced requirement to collect detailed descriptive data. The main weaknesses of the top-down approach include an over reliance on historical aggregated data which includes lack of detail for geometrical and thermal characteristics. There is also a lack of detail at individual house end use energy consumption which makes it difficult to identify specific areas which require upgrade works [18].

The bottom-up approach looks at the estimated energy consumption of a representative set of individual houses starting at a regional level and then extrapolates this data back up to measure energy consumption nationally; it consists of two methodologies: the statistical (SM) and the engineering method (EM) [19]. The EM comprises of developing a building stock database which represents the stock under analysis and then estimates the energy consumption of the stock through dynamic simulation modelling of house types or representative archetypes [20].

In contrary to top-down modelling, the bottom-up approach disaggregates components which gives an accurate description of energy usage [21]. This method however, working at a disaggregated level requires a large empirical data base to support the individual description of components [14]. Individual variables can be modified to run multiple permutations making it possible to predict specific scenarios and aid in identifying specific upgrade areas. Scenario planning using a bottom up model is referred to as a 'normative futures analyses' or 'backcasting' [18]. A study assessing available models and effects on energy saving carried out by Swan and Ugursal [19] found that a bottom-up modelling method using the EM is necessary to model impacts of new technologies.

The main characteristics of the three major residential energy modelling approaches namely top-down, bottom-up EM and bottom-up SM method are reproduced from [19] and detailed in Table 4.1.

Table 4.1 Positive and negative attributes of the three major residential energy modelling approaches [19]

	Top-down	Bottom-up statistical	Bottom-up engineering
Positive attributes	• Long term forecasting in the absence of any discontinuity	• Encompasses occupant behaviour	• Model new technologies
	• Inclusion of macroeconomic and socioeconomic effects	• Determination of typical end-use energy contribution	• "Ground-up" energy estimations
	• Simple input information	• Inclusion of macroeconomic and socioeconomic effects	• Determination of each end-use energy consumption by type, rating, etc.
	• Encompasses trends	• Uses billing data and simple survey information	• Determination of each end-use qualities base on simulation
Negative attributes	• Reliance on historical consumption information	• Multicollinearity	• Assumption of occupant behaviour and unspecified end-uses
	• No explicit representation of end-uses	• Reliance on historical consumption information	• Detailed input information
	• Coarse analysis	• Large survey sample to exploit variety	• Computationally intensive
			• No economic factors

4.3 Stock Modelling Method Used

A number of negatives and positive attributes of top-down modelling are described through the literature; and summarized above. Based on this review it was decided that the top-down approach is unfitting to the proposed method in this study. A decision was taken to apply a bottom-up approach to test a stock aggregation methodology because of the high level of detail needed. The choice of variables to describe the stock needs to be based on what outcomes are required from the study. This study is focused on developing a detailed building stock database heavily focused on geometric configuration of house types to arrive at a building model capable of accurately describing problematic areas in the thermal envelope. Therefore exposed surface areas will be weighted heaviest in the statistical composition of house types which make up representational archetypes. Vintage interval is also important in developing archetypes, as changes to thermal envelope have improved over time. There are two steps involved when using archetypes to arrive at stock aggregation, firstly, Sub totaling by multiplying results of each archetype by the amount of buildings or floor area it represents and secondly, by totalling the sub totals for each archetype to arrive at stock aggregation [18]. Figure 4.1 outlines a high level framework to arrive at stock aggregation; a combination of applying the RCM (described in Fig. 4.2) to LAH stock, to catalogue and inform the RMM used to extract building geometry and BER data for thermal characteristics. All information is aggregated to generate a LAH Building Stock Database; archetypes are then constructed and the parameters of interest modelled with the findings scaled according to the buildings they represent to arrive at stock aggregated insights.

4.3.1 Data Collection Methods

The method proposed in this study requires a detailed empirical data base of the terraced LAH stock being represented. Following a comprehensive literature review Table 4.2 was developed, which outlines the variables needed to arrive at stock aggregation. The variables are broken up into 3 categories; classification, geometrical and thermal. These three variables are discussed as follows.

4.3.1.1 Classification

The LAH stock under analysis must be catalogued and systematically disaggregated so that results from a bottom-up engineering method can be extrapolated back up to arrive at stock aggregation.

Applying a traditional data collection method can be extremely time consuming when no detailed database exists for the stock being investigated. A high speed

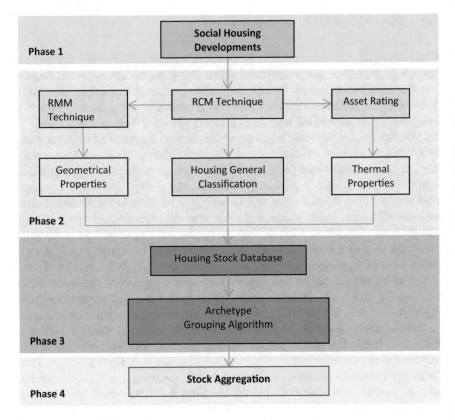

Fig. 4.1 Proposed high level framework to arrive at stock aggregation

remote method is proposed for the cataloguing of LAH stock; the Remote Cataloguing Method (RCM). The RCM is illustrated in Fig. 4.2, with a number of decision trees to define the scope of the stock being catalogued. This method facilitates rapid cataloguing of LAH stock using a combination of remote sensing images from GE and static panoramic viewpoints from GSV [22]. RCM uses a GIS based web mapping application providing high resolution satellite imagery based on location based information to identify and catalogue house types according to type h_j ($j = 1, 2,..., m$), and rank and percentiles.

4.3.1.2 Geometrical

Original construction drawings or survey information is generally used to get geometrical configuration of house types. This traditional data collection method used by Pittam et al. [23], proved to be extremely time-consuming and expensive, and was frustrated by lack of existing information. A new method is currently under development by the Authors to collect and catalogue building geometry called the

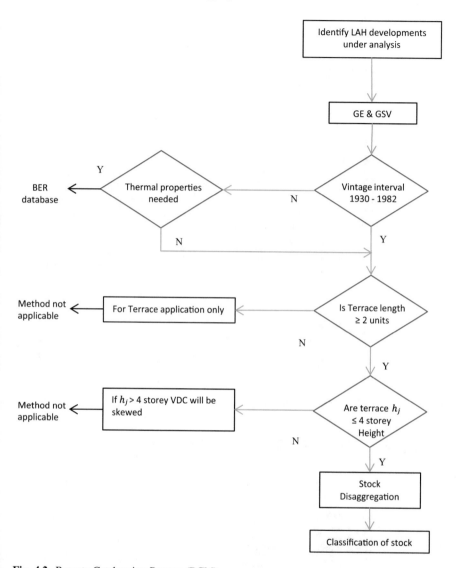

Fig. 4.2 Remote Cataloguing Process (RCM)

Remote Measurement Method (RMM). The RMM was developed as a direct result of unobtainable building specific information. The method proposed is for use on low rise domestic LAH terraced building stock. The method enables the remote extraction of building geometry from a particular terraced building type; using a combination of remote sensing images from GE and static panoramic viewpoints from GSV. This proposed method is currently submitted to a journal and is in review.

Table 4.2 Variables needed to arrive at stock aggregation

			Sources to provide variables					
			RCM	RMM	BER	S	M	CSO
Thermal	1	Wall U values			X			
	2	Roof U-value			X			
	3	Floor U-value			X			
	4	Window U-value			X			
	5	Air change rate (ac/h)				X		
	6	Internal temperature						X
	8	Heating system Eff %			X			
	9	Temperature controls			X			
	10	Number of Occupants			X			
	11	DWH cylinder insulation (mm)			X			
	12	Cylinder size (litres)			X			
	13	Pipework insulation (mm)			X			
	14	Previous upgrades (yes/no)			X			
	15	Electricity tariff rate						X
	16	Draughts			X			
	17	Humidity						X
	18	Immersion weekly frequency						X
	19	Electric shower weekly frequency						X
	20	Electric water weekly frequency						X
	21	Typical weekly occupancy patterns						X
	22	Climate data					X	
Geometrical	23	Treated floor area		X				
	24	Room Height		X				
	25	Living room area			X			
	26	Ceiling/roof area		X				
	27	External wall area		X				
	28	Window area		X				
	29	Window deviation from north		X				
	30	Window angle of inclination		X				
	31	Shading factors		X				
	32	Exposed perimeter		X				
	33	Plan layout		X				
	34	House Volume		X				
	35	Number of stories		X				

(continued)

Table 4.2 (continued)

			Sources to provide variables					
			RCM	RMM	BER	S	M	CSO
Classification	36	Dwelling type	X					
	37	Number of terrace typologies	X					
	38	Number of units per typology	X					
	39	Categorize MT, ET, SD	X					
	40	Terrace orientation	X					
	41	Terrace elevation	X					
	42	Categorize by construction type	X					

S Sinnnot, *M* Meteonorm, *CSO* Central Statistics Office

4.3.1.3 Thermal

Thermal characteristics are defined from published BER data for ET and MT units carried out on Cork City terraced housing stock; using the DEAP software program, which is based on the calculation framework I.S. EN ISO 13790: 2004: Thermal performance of buildings [24]. It is very closely aligned with the UK's calculation method, the Standard Assessment Procedure (SAP) [25]. DEAP is Ireland's official BER tool used to show compliance with building regulations and the methodology framework in the Energy Performance of Buildings Directive (EPBD) [24]. Data from compiled results is broken up into 481 ET BER's and 1073 MT BER's; it is then subdivided up into DEAP vintage categories ranging from 1930 to 1949; 1950 to 1966; 1967 to 1977 and 1978 to 1982. Thermal characteristics need to include envelope components, thermal insulation levels, type of windows, heating and hot water efficiencies, equipment data and air tightness data. Table 4.2 describes thermal variables needed and sources to provide them. 23 key variables are identified in Table 4.2 from a study carried on by Famuyibo et al. [26], with the remaining variables identified through a composite of published literature.

4.4 Methodology for Archetype Development

As mentioned above there are a number of different methodologies currently documented using stock aggregation techniques for the development of virtual archetypes for use in housing stock energy modelling. Table 4.3 outlines the methodology that was developed for this study.

Each step is identified separately with an attempt to provide a description of the general principle applied. The particular solution satisfying each step for this study is summarised separately. Broadly the method currently relies on a bottom up EM approach where its strengths are dependent on availability of certain types of data from local or national authorities. Based on the above methodology, archetype

Table 4.3 Stock aggregation and archetype development methodology

	Description	Comments from this study
1	Identify total LAH housing stock, S_h, based on local council planning boundaries	City map data obtained from council housing department used to define scope. Refer to Fig. 4.3
2	Use a GIS based web mapping application providing high resolution satellite imagery based on location based information to identify and catalogue house types according to type, $h_j(j = 1, 2, ..., m)$ and rank and percentiles	Google street view web based GIS system used with 18 house types defined from a total of 10,318 houses across 36 developments
3	Remove any house types that have percentile values below a selected threshold value or can be discounted based on criteria set out in the stock aggregation model rules	In this instance we applied an exclusion criteria based on construction type rather than threshold quantities. Red brick properties excluded based on delineation of retrofit solutions
4	Record/Calculate all geometrical configurations and non-geometric (thermal, electrical) characteristics for each house type, h_j, using available sources	In this instance data was obtained using the RMM and results from a national asset rating study of approximately 1551 houses (see Tables 4.4 and 4.5)
5	Group house types, h_j, according to construction type as a 1st tier grouping rule and according to surface volume, SV, as a 2nd tier grouping rule	18 house types were combined to give 4 archetypes. See Fig. 4.6
6	Define archetype column vector, a_i, ($i = 1, 2, ..., n$) based on parameter weighting Eq. (1) for each parameter p_k ($k = 1, 2, ..., l$) using w_j	Parameters identified in Table 4.2 were used in the column vector a_i

parameter column vectors derived in step 6 are based on the product matrix in Eq. (4.1). This column vector contains all parameters describing the archetype such as u values, window areas, roof areas, airtightness performance, boiler efficiency etc. Each parameter, $p_{k,i}$, within archetype dataset $\boldsymbol{a_i}$, where i represents the individual archetype, is defined as:

$$p_{k,i} = \sum_{j=1}^{m} w_j p_{kj} \tag{4.1}$$

Weighting factors, w_j, are based on the proportional contribution from different house types to each archetype parameter in the column vector and can be defined as:

$$w_j = \frac{1}{\sum_{j=1}^{m} X_i} X_j \tag{4.2}$$

where $X_j \in S_h$ and is the total number of units for house type h_j. In general it is possible to produce each archetype parameter dataset, $a_i = [p_{1,i}, p_{2,i}, ..., p_{k,i}]$, using the matrix form of a linear system as a single vector equation defined in (4.3) as:

$$[p|w]_i = a_i \tag{4.3}$$

where the coefficients of Eq. (4.3) will be different for each archetype based on the sub classification of house types and can be defined using the $l \times m$ matrix P_i and the two column vectors w_i and a_i as shown in (4.4).

$$p_i = \begin{bmatrix} p_{1,1} & p_{1,2} & \cdots & p_{1j} & p_{1,m} \\ p_{2,1} & p_{2,2} & \cdots & p_{1j} & p_{2,m} \\ \cdot & \cdot & \cdots & \cdot & \cdot \\ p_{k,1} & p_{k,1} & \cdots & p_{k,1} & p_{k,m} \\ p_{l,1} & p_{l,2} & \cdots & p_{l,j} & p_{l,m} \end{bmatrix}, w_i = \begin{bmatrix} w_1 \\ w_2 \\ \vdots \\ w_j \\ w_m \end{bmatrix}, a_i = \begin{bmatrix} p_{1,i} \\ p_{2,i} \\ \vdots \\ p_{k,i} \\ p_{l,i} \end{bmatrix} \tag{4.4}$$

In summary the method relies on working closely with national authorities that have certain types of data available. The use of GSV is a key element of the method as it facilitates remote typology cataloguing of a large amount of building stock in a short period of time using a low cost approach.

4.5 Application of Stock Aggregation Method

This section describes the application of the stock aggregation method described above to the Cork City LAH stock to build a detailed building stock database containing classification of stock, geometric configuration and thermal character-istics. The first step was to Identify total LAH housing stock, S_h, based on local council planning boundaries. Figure 4.3 presents a map of Cork city with 36 LAH developments highlighted.

The RCM uses Google Earth and Google street view to collect and catalogue information on LAH stock under analysis, Fig. 4.4 illustrates an example of this. This high speed remote cataloguing approach enabled a full systematic classifica-tion and disaggregation of LAH stock within a number of weeks.

By applying the RCM, the distribution of terraced stock across each LAH development can be analysed and relationships between stock formations evaluated. Figure 4.5 illustrates an example of the application of the RCM to Ballyphehane housing development number 9 (see Fig. 4.3 for position on map). Each h_j is colour coded to the various terrace lengths, so distribution of h_j and t_l can be uploaded to the building stock database.

There are a total of 1570 units in development 9 (Fig. 4.5), made up of house types h_1, h_6, h_8, h_{13}, h_{14} and h_{16}. Within development 9, h_1 makes up 68% of total number of units, and is a composite of $56 \times 2t_l$, $87 \times 4t_l$, $52 \times 6t_l$, $31 \times 8t_l$ and

Fig. 4.3 Map of Cork cities LAH stock

Gurranabraher development 22   Togher development 8

Fig. 4.4 RCM using GE and GSV

$4 \times 10t_l$ variations. The roof area potential to mount solar PV or solar thermal for h_1 is broken up into South West, South and South East amounting to 760, 3940 and 1866 m^2 respectively. There is total of 18020 m^2 of glazing area for terrace house type h_1 in development 9; cost estimates can be made remotely using this database and the impact of retrofitting individual components analysed through dynamic simulation modelling. House type, h_1 is of block on flat construction and is the most repeating house type across all 36 developments amounting to 2160 units. Wall area

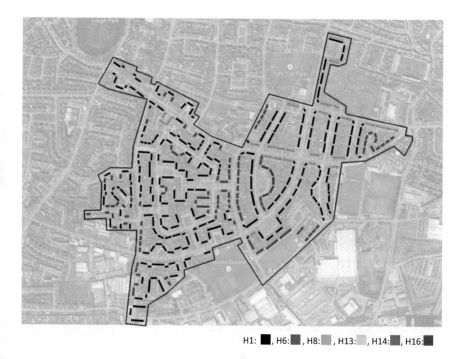

H1: ▇, H6: ▇, H8: ▇, H13: ▢, H14: ▇, H16: ▇

Fig. 4.5 Example of RCM applied to LAH development 9

potential for an external insulation application can also be assessed at development level and energy savings modelled. This level of detailed analysis is carried out across all 36 LAH developments and outlines the importance of going to this level of stock disaggregation.

From the evaluation of the 36 LAH developments 18 repeating terraced house variations are recorded and presented in Fig. 4.6, with total number of units, denoted S_h amounting to 10,318. The remaining units excluded based on delineation of retrofit solution are made up of 5% red brick and 2% outlier units from the original 11,029 units. Any house types that have percentile values below a selected threshold value or can be discounted based on criteria set out in the stock aggregation model rules are removed. In this instance we applied an exclusion criteria based on construction type rather than threshold quantities. Red brick properties excluded based on delineation of retrofit solutions. Data is broken down into MT, ET and SD units, amounting to 55, 32 and 13% respectively. Total percentage orientation for North, South, West and East is 29, 27, 23 and 21% correspondingly.

The distribution of stock across the 36 LAH developments is outlined in Fig. 4.7, with development number 9 highlighted in red which is used as an example to illustrate RCM in Fig. 4.5. Development numbers in Fig. 4.7 display results from numbered developments on map in Fig. 4.3.

Fig. 4.6 House types, h for 10,318 dwellings across 36 developments

The distribution of total number of house types, h_j across all 36 developments is presented in Fig. 4.8. House types are further distributed across each development as seen in Fig. 4.5, which makes large scale optimum decision making to evaluate upgrade works possible using the housing stock database.

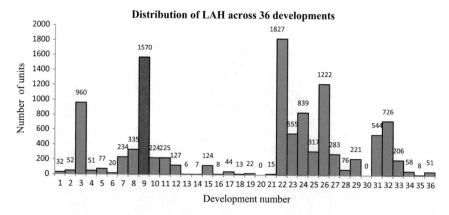

Fig. 4.7 Distribution of LAH stock across 36 housing developments

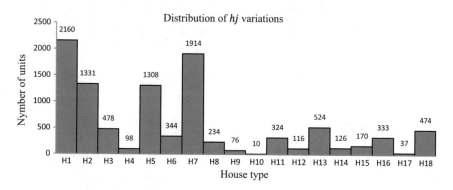

Fig. 4.8 Distribution of house types, h_j

4.5.1 Results for Geometrical and Thermal Characteristics

Terrace lengths are initially calculated using GE and GSV. GSV is used to identify h_j and GE is used to measure roof length and width using the size capture function tool illustrated in Fig. 4.9.

The number of terraces for each t_l variation for all h_j, are outlined in Table 4.4. This information is extremely important when applying a stock aggregation methodology. Each h_j is analysed for MT and ET with geometrical and thermal characteristics extracted from available sources of data. To arrive at stock aggregation the database must contain total number of units for each h_j and a disaggregated composition of each t_l variation which make up that number. Results from modelling process can then be multiplied by total number of units based on type for each t_l variation, to assess options at terrace level; or alternatively by total number of units broken up into ET and MT to arrive at stock aggregation. By having a

Google Earth

Roof Length measured using size capture function from RS view

Fig. 4.9 Size capture function tool in GE

Table 4.4 Distribution of t_l across 36 LAH developments for 18 h_j

h_j	$2t_l$	$3t_l$	$4t_l$	$5t_l$	$6t_l$	$7t_l$	$8t_l$	$9t_l$	$10t_l$	$12t_l$
h_1	126		179		102		57		17	
h_2	62		50		59		49		27	
h_3	10		9		28		16		7	
h_4	3		3		4		7		1	
h_5					218					
h_6	203	26	30		10		1			
h_7	72	10	144	15	108	6	30		8	
h_8	40	3	22	4	5					
h_9			3		4		3		1	
h_{10}	5									
h_{11}	5	5	8	5	16	6	5	4	4	
h_{12}	4	3	2	1	2	8	1		1	
h_{13}	8	3	44		55					
h_{14}	43		5							
h_{15}					5		2		11	1
h_{16}		2	1		11			16	7	5
h_{17}		1	1		2		1		1	
h_{18}	49	22	56	6	7		2			

Table 4.5 Built form characteristics from RMM application

	h_1	h_2	h_3	h_4	h_5	h_6	h_7	h_8	h_9	h_{10}
V (m³)	150.50	155.20	149.60	148.90	212.20	127.50	197.00	166.00	213.70	242.00
GF A (m²)	31.00	31.20	31.20	33.80	38.60	54.50	40.20	33.20	41.10	46.50
S/V MT	0.80	0.70	0.70	0.80	0.70	1.10	0.70	0.90	0.70	0.70
S/V ET	1.02	0.93	0.94	0.96	0.85	1.28	0.87	1.01	0.85	0.86
P/A MT	0.30	0.30	0.30	0.30	0.30	0.29	0.26	0.44	0.33	0.33
P/A ET	0.51	0.51	0.51	0.50	0.47	0.42	0.45	0.58	0.48	0.46

detailed disaggregated geometrical and thermal database it offers a much deeper analysis into various options to improve thermal performance at a local or national level. With this high level of information, detailed high impact retrofit permutations can be analysed at a terrace level to develop cost optimum terrace level solutions.

All geometrical variables highlighted in Table 4.5 are extracted using the RMM on the 18 terraced house types identified in Fig. 4.6. Table 4.5 presents an example of results from 10 house types using the RMM to generate the geometrical configuration of each house type subdividing data into ET and MT.

Thermal characteristics and energy rating averages for 1551 BER's for Cork are described in Table 4.6. The data is broken up into vintage intervals ranging from 1930 to 1982 to assess the variations in energy rating between vintage periods for MT and ET. Vintage intervals range from 1930 to 1949, 1950 to 1966, 1967 to 1977 and 1978 to 1982 with variations in energy rating between MT and ET being 11, 11, 13 and 9% respectively. ET typologies have an averaged 11% higher energy rating overall partly due to increased wall area described in Table 4.3; also floor U-values for vintage intervals are 20, 9, 23 and 22% higher because of increased perimeter to area ratio in ET typologies. Wall U-values for MT and ET from 1930 to 1982 improve by 57 and 65% respectively having a major impact on the energy rating. DHW cylinder insulation thickness for ET and MT from 1930 to 1982 increases by 36 and 21%. The overall energy rating for MT and ET improves by 37 and 38% correspondingly from 1930 to 1982. From aggregated interval survey data it can be seen in Table 4.6 that space heating boilers have been upgraded with space heating efficiency being above 80% across the survey sample. Window U-values are very poor across each vintage interval and an upgrade to this area would have a major impact on improving building energy performance and thermal comfort.

4.6   Archetypes Development

This section presents the geometry-oriented archetypes with geometry built from a composite of data extracted from the RMM and from aggregated survey data from a national asset rating study of 1551 BER's from Cork cities LAH terraced stock. Figure 4.10 describes archetype development and extrapolation of data to arrive at

Table 4.6 Aggregated thermal values in vintage intervals

Vintage	T	N	Energy rating (kWh/m²/year)	CO_2	Space heating (Eff %)	Floor (W/m²K)	Wall (W/m²K)	Roof (W/m²K)	Win (W/m²K)	DWH cylinder (mm)
1930–1949	MT	195	326.99	70.99	81.61	0.61	1.34	0.86	3.21	19.99
1930–1949	ET	71	366.53	84.76	84.06	0.77	1.42	0.80	3.30	23.94
1950–1966	MT	316	331.75	72.52	81.75	0.68	1.56	0.66	3.13	25.62
1950–1966	ET	146	374.11	83.66	80.03	0.75	1.57	0.73	3.26	28.16
1967–1977	MT	370	232.94	49.65	82.87	0.60	0.73	0.40	3.05	27.30
1967–1977	ET	176	266.58	57.97	83.49	0.78	0.65	0.44	3.11	26.39
1978–1982	MT	191	206.13	43.86	82.45	0.47	0.57	0.25	2.90	31.26
1978–1982	ET	86	227.47	48.87	84.11	0.60	0.49	0.25	2.90	30.35

N Sample distribution, *T* Type), *Win* Window

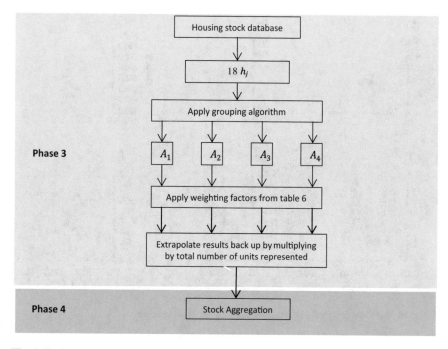

Fig. 4.10 Archetype development and results extrapolated to arrive at stock aggregation

stock aggregation. Phase 3 from Fig. 4.1 is further expanded in Fig. 4.10. Thermal data is taken directly from the aggregated survey results and grouped into vintage intervals as described in Table 4.6.

A number of geometrical variables are graphed in Fig. 4.11 for each of the 18 h_j against floor area. From the analysis of data; Construction type is used as a 1st tier grouping rule with Height, Volume and Floor area used as a 2nd tier grouping rule which is also compared to survey BER data to ensures transparency between methods. Graph (f) in Fig. 4.11 shows width variations with 1 storey having wider width and 3 storey narrowest. Graph (b) shows terrace depth range to be quiet similar for over 83% of h_j. Graph (d), and Graph (e) group h_j into 1S, 2S and 3S.

House types, h_j for 2S dwellings are subdivided into two vintage categories to optimize the accuracy of results. This also divides block on flat constructed house types, h_j and cavity constructed h_j's. Thermal characteristics from survey data are then grouped to each geometrical grouping based on vintage interval. Weighting factors are applied to the parameters defining each dwelling geometrical and thermal characteristic's and are presented in Table 4.7. Table 4.7 also highlights the distribution of house types which are aggregated to create each representative archetype. The weighting of variables ensures that archetypes are weighted based on individual house type contribution subdivided into ET and MT; giving us an accurate description of the LAH stock under analysis and ensures representative results are extrapolated from the energy modelling of archetypes.

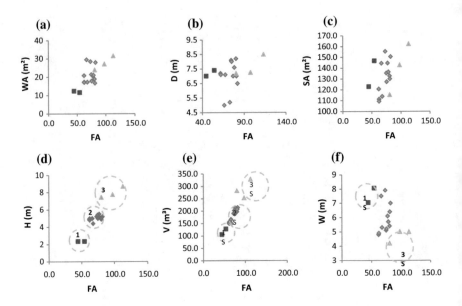

Fig. 4.11 variables graphed to establish relationships in H^t grouping

Airtightness data used for each archetype is taken from a study carried out by Sinnot et al. [27], on the Irish housing stock. Air change rate (ac/h) are based on mean air permeability tests carried out on Irish domestic properties built between 1941 and 1975 and from 1980 to 1986; results for tests are 7.5 and 9.4 $m^3/h/m^2$ respectively [27].

4.7   Results from Case Study

Once the archetypes were defined it was possible to do a base line energy assessment by modelling each archetype using DEAP software program, Table 4.8 summarizes the archetype geometrical column vectors into sub groups, ET and MT. Figure 4.12 illustrates two 3D Sketch-up models which represent Archetype A_2 and A_3. The models are built based on the weighting of geometrical variables from the terrace h_j they represent (see Table 4.8). 2 archetypal sections for archetype A_2 and A_3 are also illustrated in Fig. 4.12. Sketch-up 3D models are generated from extracted 2D information during RMM process. Sketch-up can be plugged into a number of dynamic simulation packages for energy modelling purposes. An extension of the research work presented here will be focusing on modelling multiple retrofit permutations using the sketch-up archetypal models. Table 4.8 describes geometrical variables for A_2 and A_3 weighted based on house type contribution. Terrace length variations 2 t_l, 4 t_l, 6 t_l and 8 t_l are presented in Table 4.8 which demonstrates the disaggregation of built form characteristics extracted using the RMM. Thermal

Table 4.7 Weighting factors applied to each house type and archetype grouping

a_i		ET								MT						
A_1	h_j	h6a	h6b							h6(a)	h6(b)					
	w_j	0.5	0.5							0.5	0.5					
A_2	h_j	h1	h2	h3	h4	h8	h14			h1	h2	h3	h4	h8	h14	
	w_j	0.49	0.29	0.07	0.02	0.08	0.05			0.48	0.31	0.14	0.03	0.03	0.004	
A_3	h_j	h5	h7	h9	h10	h11	h12	h13	h18	h5	h7	h9	h11	h12	h13	h18
	w_j	0.22	.41	0.01	0.01	0.06	0.02	0.11	0.15	0.31	0.39	0.02	0.07	0.03	0.11	0.06
A_4	h_j	h15	h	h17						h15	h16	h17				
	w_j	0.29	0.67	0.05						0.35	0.56	0.09				

Table 4.8 Variables weighted for Archetype A_2 and A_3

	A_2 Block on flat construction				A_3 Cavity construction			
	$2t_l$	$4\,t_l$	$6\,t_l$	$8\,t_l$	$2\,t_l$	$4\,t_l$	$6\,t_l$	$8\,t_l$
l (m)	12.2	24.3	33.9	44.9	11.8	22.9	35.5	46.6
d (m)	6.4	6.6	6.7	6.7	7.6	7.5	7.0	7.7
h (m)	4.9	4.9	4.9	4.9	5.1	5.1	5.3	5.0
V (m³)	383.0	777.6	1107.4	1476.5	456.3	876.3	1303.2	1777.8
win/door (m²)	29.0	57.6	82.9	108.7	27.9	54.6	110.2	107.9
P (m)	35.6	55.0	73.8	92.9	38.4	60.1	84.7	103.9
P/A ratio	0.5	0.4	0.4	0.4	0.5	0.4	0.4	0.3
WA (m²)	153.0	241.8	311.6	384.0	168.5	249.6	338.8	412.5
GF A (m²)	65.6	125.6	186.9	249.7	80.4	160.6	234.6	319.3
S/V ratio	0.82	0.71	0.69	0.79	0.79	0.72	0.71	0.65

P Perimeter, *A* Area, *W* Wall, *GF* Ground-Floor

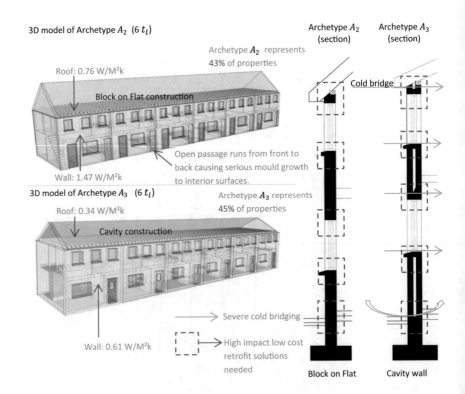

Fig. 4.12 3D model and sectional wall detail for archetype B and C

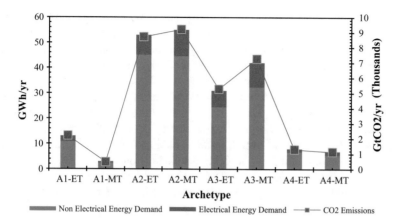

Fig. 4.13 Summary of base line energy assessment

characteristics for archetypes are extracted from Table 4.6 and weighted based on vintage category linked to each house type.

Figure 4.13 summarizes the base line energy assessment showing DEAP simulated results. The total end use energy consumption and CO_2 for each archetype are displayed in the graph. Results are based on geometrical data from Table 4.8 and thermal data from Table 4.6. Percentage thermal and electrical energy use is subdivided to demonstrate where potential savings can be made. A full estimation of total energy use and CO_2 is recorded for 10,318 units resulting in 211 GWh/y and 35,477 $GtCO_2$/year.

4.8 Discussion

A detailed housing data base for city council LAH was not in existence and was necessary in order to carry out a study of this nature. Using a traditional approach this would have been unfeasible but applying a web based GIS mapping approach to rapid typology cataloguing reduced substantially the time taken to develop such a database. As the second part of this study will be focusing on large scale external retrofit solutions it necessitates an accurate subdivision and disaggregation of geometrical and thermal data. Results from this study are compared to a recent study carried out by Famuyibo et al. [26] where they developed 13 archetypes to represent the Irish housing stock more generally. Floor, roof and wall U-values are quite comparable with their floor, roof and wall U values ranging from 0.50 to 0.58, 0.33 to 0.46 and 0.50 to 1.63 W/m²k respectively. The floor, roof and wall U value results from this study range from 0.47 to 0.78, 0.25 to 0.86 and 0.49 to 1.57 W/m²k respectively. The largest variance is in the floor areas where Famuyibo et al. archetypes are all over 100 m² although this is expected due to large volume of house typologies used in their study [26]. The study house types used by Sinnot et al. [27] to measure air permeability have an average floor area and internal volume of 80 m^2 and 202 m^3 respectively. In this study archetype average floor area and volume are 83 m^2 and 196 m^3 respectively. They recorded an average 5% reduction compared to an average 2.46% reduction in theoretical energy consumption in this study when using Sinnots results. These reductions may appear minor but are extremely important in increasing the accuracy between estimating simulated energy consumption and actual energy consumption of existing dwellings. That 2.46% if added to overall results would amount to 873 GtCO$_2$/year. This study is a step in contributing to a highly detailed national housing database; which could be used for cost optimal retrofit decision making and to do accurate energy simulations and assess multiple retrofit scenarios in a bid to help reduce carbon emissions and bridge the gap in meeting our Kyoto obligations.

4.9   Conclusion and Future Work

This paper outlines a method for cataloguing LAH in Cork City using a combination of GIS, a web mapping application, RMM and survey data to build a detailed housing database. From the collection of data, archetypal buildings are produced using a matrix based linear system which combines a series of house types and parameters into archetypes to execute large scale energy modelling analysis. The method used to apply weighting factors to building variables based on unit quantity proved to be an effective measure in creating accurate archetypes. This leads to a more representative level of accuracy in results from energy modelling. It is proposed to use a similar method to increase accuracy in U value calculation, as certain construction elements have a combined material build up.

In conclusion the method has produced a series of archetypes; the authors believe the 4 principle archetypes developed are representative of LAH stock in Cork City. The 4 archetypes represent 94% of the original sample stock.

The method used was highly effective in collecting the data needed to build a detailed housing data base for Cork City. It is believed that this method is adaptable and could be successfully applied to other city council housing stocks.

References

1. Pulselli, R., et al.: Emergy analysis of building manufacturing, maintenance and use: Embuilding indices to evaluate housing sustainability. Energy Build. **39**(5), 620–628 (2007)
2. SEAI, Energy in the Residential Sector, 2013 Report, Ireland
3. Brophy, V., et al.: Homes for the 21st century: the costs & benefits of comfortable housing for Ireland. Energy Action Limited (1999)

4. Lapillonne, B., Sebi, C., Pollier, K.: Energy efficiency trends for households in the EU. Enerdata—an analysis based on the ODYSSEE Database (2012)
5. SEAI: Energy in the residential sector. In: Ireland, S. E. (ED), Dublin (2008)
6. Clinch, J.P., Healy, J.D.: Housing standards and excess winter mortality. J. Epidemiol. Commun. Health **54**(9), 719–720 (2000)
7. Healy, J.D., Clinch, J.P.: Quantifying the severity of fuel poverty, its relationship with poor housing and reasons for non-investment in energy-saving measures in Ireland. Energy Policy **32**(2), 207–220 (2004)
8. SEAI: A review of fuel poverty and low income housing (2003)
9. Marcin Badurek, M.H., Bill, S.: Building typology Brochuree Ireland—a detailed study on the energy performance of typical Irish dwellings (2012)
10. Lechtenböhmer, S., Schüring, A.: The potential for large-scale savings from insulating residential buildings in the EU. Energ. Effic. **4**(2), 257–270 (2011)
11. Johnston, D.: A physically-based energy and carbon dioxide emission model of the UK housing stock. Leeds Metropolitan University (2003)
12. Firth, S.K., Lomas, K.J., Wright, A.: Targeting household energy-efficiency measures using sensitivity analysis. Build. Res. Inf. **38**(1), 25–41 (2010)
13. Shimoda, Y., et al.: Residential end-use energy simulation at city scale. Build. Environ. **39**(8), 959–967 (2004)
14. Shorrock, L., Dunster, J.: The physically-based model BREHOMES and its use in deriving scenarios for the energy use and carbon dioxide emissions of the UK housing stock. Energy Policy **25**(12), 1027–1037 (1997)
15. Parekh, A.: Development of archetypes of building characteristics libraries for simplified energy use evaluation of houses. In: Proceedings of 9th International IBPSA Conference (2005)
16. Petersdorff, C., Boermans, T., Harnisch, J.: Mitigation of CO_2 emissions from the EU-15 building stock. beyond the EU directive on the energy performance of buildings. Environ. Sci. Pollut. Res. **13**(5), 350–358 (2006)
17. Swan, L., Ugursal, V.I., Beausoleil-Morrison, I.: A new hybrid end-use energy and emissions model of the Canadian housing stock. In: The First International Conference on Building Energy and Environment COBEE, July 2008
18. Moffatt, S.: Stock aggregation: methods for evaluating the environmental performance of building stocks. Annex 31, Energy-Related Environmental Imapacts of Buildings (2004)
19. Swan, L.G., Ismet Ugursal, V.: Modeling of end-use energy consumption in the residential sector: a review of modeling techniques. Renew. Sustain. Energy Rev. (2008)
20. Aydinalp, M., Ugursal, V., Fung, A.: Modelling of residential energy consumption at the national level. Int. J. Energy Res. **27**(4), 441–453 (2003)
21. Kavgic, M., et al.: A review of bottom-up building stock models for energy consumption in the residential sector. Build. Environ. **45**(7), 1683–1697 (2010)
22. Haala, N., Kada, M.: An update on automatic 3D building reconstruction. ISPRS J. Photogrammetry Remote Sens. **65**(6), 570–580 (2010)
23. Pittam, J., O'Sullivan, P.D., O'Sullivan, G.: Stock aggregation model and virtual arche-type for large scale retro-fit modelling of local authority housing in Ireland. Energy Procedia **62**, 704–713 (2014)
24. SEAI: Dwelling energy assessment procedure (DEAP) VERSION 3.2.1. Irish official method for calculating and rating the energy performance of dwellings, sustainable energy authority of Ireland (2012)
25. SAP: The government's standard assessment procedure for energy rating of dwellings. Version 9.92, Oct 2012
26. Adesoji Alvert Famuyibo, A.D., Paul S.: Developing archetypes for domestic dwellings—an Irish case study. Energy Build. (2012)
27. Sinnott, D., Dyer, M.: Air-tightness field data for dwellings in Ireland. Build. Environ. **51**, 269–275 (2012)

Chapter 5
Renewable Homes—Feasibility Options for UK Traditional Buildings Through Green Deal

Charikleia Moschou and Catalina Spataru

Abstract UK government set up several policies to promote renewable technologies uptake and to improve the energy efficiency of dwellings. Green Deal is the most recent UK policy and it intends to promote the energy efficient refurbishment of existing dwellings. This study explores potential packages through Green Deal for five different dwelling types. The proposed packages include: the use of insulation, glazing and renewable energy technologies. This paper proposes three potential packages for each dwelling type, which would have the ability to offer the maximum amount of CO_2 savings. A model has been developed in order to evaluate the recommended packages. It seems that the package of: internal, cavity, floor, roof/loft, doors insulation and micro turbine could offer the most significant CO_2 reduction in detached dwellings, while the package of: external, cavity, floor, roof/loft, doors insulation and a wind turbine could offer the lowest amount of CO_2 savings in terraced dwellings.

Keywords Buildings Green Deal Energy modeling Renewables Technologies

5.1 Introduction

UK buildings account for the 45% of total UK carbon emissions [1]. According to studies, approximately 29% of carbon dioxide emissions corresponds to the domestic sector [2]. Generally, UK housing stock consists of 26 million dwellings and almost 9.2 million dwellings are considered as hard-to-treat homes [1]. UK government has set out environmental policies to meet the national targets for the

C. Moschou (✉)
Department of Civil, Environment and Geomatic Engineering,
UCL, Gower Street, London WC1E 6BT, UK
e-mail: c.moschou.12@ucl.ac.uk

C. Spataru
Energy Institute, UCL, Central House, 14 Upper Woburn Place,
London WC1H 0NN, UK

© Springer International Publishing AG 2017
J. Littlewood et al. (eds.), *Smart Energy Control Systems for Sustainable Buildings*,
Smart Innovation, Systems and Technologies 67,
DOI 10.1007/978-3-319-52076-6_5

carbon reduction. Green Deal is the most recent environmental policy in UK and its purpose is to pro- mote the improvement of the energy efficiency for the existing UK building stock.

From January to December 2013, the statistics revealed that 117,454 physical assessments of dwellings were carried out, which reflects that the penetration rate is almost 0.5% of the UK housing stock [3]. It seems that, there are concerns that Green Deal will fail to deliver the required carbon cuts to meet UK's emissions reduction targets. Even though the advantages of Green Deal are multiple, the adoption rates of the implementation of Green Deal packages are significantly slow [3].

This paper will review the potential reduction in CO_2 emissions under different packages as part of the Green Deal, by exploring various options of Green Deal and also the use of energy saving techniques that can be exercised within existing homes to reduce energy use and carbon emissions. It is necessary to examine different potential Green Deal packages and to evaluate the CO_2 reduction that they can offer to occupants. The particular study intends to conduct a practical assessment perspective of these potential packages and also to establish "ideal" packages for five different dwelling types. The proposed Green Deal packages will offer the maximum amount of carbon dioxide reduction, they will increase the energy efficiency of dwellings and at the same time they should be cost effective.

5.2   Green Deal

The main purpose of Green Deal is to improve the energy efficiency of UK building stock and also to achieve a significant reduction in the amount of CO_2 emissions [4]. According to the mechanism of Green Deal, the consumers can install specific energy saving measures to their properties without paying for the upfront cost of refurbishment [5]. At the same time, another advantage of this environmental policy is the fact that the implementation of Green Deal packages can offer an increased reduction in the amount of energy bills. In addition, the installation of energy saving measures, which are included to Green Deal, contributes to the limitation of the phenomenon of draught and condensation, as well as, it can increase the property value [3].

The "golden rule" of Green Deal is the criterion to determine if a Green Deal package can be financed or not. According to the golden rule of Green Deal, the expected savings, which are derived from a Green Deal package, must be equal or greater than the annual repayment costs [6]. This limitation intends to allow householders achieve savings on energy bills as soon as the implementation of the package is needed [4]. The main difference between the mechanism of Green Deal and a loan is that the payment of Green Deal is attached to the property, while the payment of loan is attached to the bill payer. The main purpose of golden rule is to secure that both customers and investors do not take financial risks [5].

Table 5.1 Phases for Green Deal finance

Phases for Green Deal finance	
First phase	The potential measures of Green Deal must meet eligibility criteria in secondary legislation
Second phase	A physical assessment of the building is conducted. A Green Deal assessor makes recommendations about the proposed energy saving measures, which are suitable for the property
Third phase	The proposed package must meet the Golden Rule of Green Deal, while the installed technologies must be in compliance to safety, health and performance standards, which are included to the Green Deal code of practice and the standards of European and domestic legislation about buildings [6]

In other words, the golden rule of Green Deal intends to check if the duration of the payment period is equal to the lifetime of the installed technologies or equal to a specific payback period. The particular limitation is only a guideline and not a rule and it intends to ensure that occupants will enjoy the advantages of the retrofit without having to pay additional net costs [7].

Table 5.1 presents the three basic steps in order to ensure that a Green Deal package can be financed. According to the first phase, specific principles of UK legislation set rules about the eligibility of Green Deal finance. The potential energy saving measures of Green Deal must not be portable. In other words, the application of a Green Deal package must be attached to a particular property, instead of being linked to an owner or a tenant. A Green Deal provider should prepare an appropriate plan for the implementation of a Green Deal package and he must have consent by the energy bill payer and all the relevant parties [6].

Then, an accredited advisor conducts a physical assessment of the dwelling [6]. The assessment determines the potential amount of savings in energy cost [3]. The accredited advisor proposes specific energy saving measures, which are suitable for the particular property. The aforementioned assessment should include an updated version of the Energy Performance Certificate (EPC) [6]. The Green Deal package consists of the recommended energy saving measures, which are proposed by the advisor. In other words, the advisor should recommend the "ideal" Green Deal package that could improve the environmental behavior of the building and could offer smart and energy saving solutions [5, 7]. The energy saving measures depend on the characteristics, the current situation and the location of the building.

In the third phase of Green Deal finance, it is necessary to secure that the recommended package can meet the golden rule of Green Deal. The Green Deal provider is responsible for preparing a particular offer, which is based on the physical assessment of the second phase of Green Deal finance. In addition, the Green Deal provider should estimate the total installation cost of the proposed measures, which can cover the energy needs of the building. Another aspect of the third phase of Green Deal finance is that the recommended energy saving technologies must comply with safety, health and performance standards of Green Deal

Code of Practice. They must also meet the standards of European legislation about the domestic sector. Moreover, the Green Deal provider should predict and evaluate if the recommended Green Deal package could meet the golden rule of Green Deal. This means that, in the third phase of Green Deal finance, it is necessary to secure that the energy savings of the Green Deal package will be equal or greater than the repayment cost. If this restriction is satisfied, then the recommended package can be implemented and financed [6].

It seems that the purpose of Green Deal is to make an effort to support the application of energy efficient solutions for domestic purposes; solutions which can achieve increased savings in the energy bills of occupants. The launch of Green Deal was considered as "a green revolution", as it promises that the annual savings in fuel bills could be equal to the amount of £550 per house. According to this prediction, it is estimated that the savings which are derived from the potential reduction of gas consumption, will be higher than £2.5 billion.

Even if the benefits of Green Deal are multiple and promising, there are concerns that the challenges and the goals of the Green Deal scheme cannot be achieved easily. Potential barriers of Green Deal are:

- Lack of consumer awareness and appeal

The Green Deal packages do not contain grants for retrofit works, which means that there is the possibility of low consumer interest about the installation of energy saving measures [5]. Additional barriers are the lack of awareness and high or potentially unrealistic expectations of consumers [3].

- Lack of investor incentives

In terms of investors, the implementation of Green Deal is not promoted as a way to gain money. For some people, this may mean that the application of a Green Deal package may be more attractive to assessors, installers and financial providers than to investors.

- Technical barriers

It seems that households, which deal problems of fuel poverty, are ignored by Green Deal. These households are estimated almost a fifth of the total housing stock. If the occupants of a house struggle to pay the energy bills for the fuel consumption, it will not be easy for them to cover the needs of a Green Deal package. As it is has already been mentioned, one of the innovations of Green Deal is the fact that the package is not attached to the occupant, but to the property. The market applications of this innovation are not yet tested [5].

Further barriers may be the limited access to information about the mechanism of Green Deal, the lack of trust in contractors, the lack of financing, the potential uncertainty of customers and complexity issues [3].

The potential energy-saving measures, which are included to Green Deal, are shown at Table 5.2 [6].

Table 5.2 Energy saving improvements of Green Deal

Heating, ventilation and air conditioning	• Heating controls • Under-floor heating • Radiant heating • District heating • Biomass room heaters • Condensing boilers (gas or oil) • Warm-air heating • Heat recovery systems • Shower heat recovery • Mechanical ventilation (non-domestic) • Flue gas recovery devices	Lighting	• Lighting fittings • Lighting controls
		Water heating	• Innovative hot water systems • Solar water heating • Water efficient taps and showers • Hot water controls
Building fabric	• Cavity wall insulation • Loft insulation • Flat roof insulation • Internal wall insulation • External wall insulation • Duct insulation • Draught • Proofing floor insulation • Heating system insulation • Energy efficient glazing and doors • Double glazed windows • Insulated doors	Microgeneration	• Ground source heat pumps • Air source heat pumps • Water source heat pumps • Solar thermal • Solar PV • Biomass boilers • Micro-CHP • Micro wind generations • Transpired solar collectors

Generally, the most effective energy saving measures, which can achieve a remarkable reduction in the amount of energy consumption and CO_2 emissions in the domestic sector, are: Insulation, Glazing, Ventilation and Micro-Distributed Technologies. These are discussed in the following sections.

5.2.1 Insulation

The use of insulation is one of the most effective ways to improve the energy efficiency in the domestic sector. Insulation systems can achieve a remarkable reduction of energy consumption, because they can keep the heat in the internal environment of a house during winter days. Respectively, during the summer, insulation systems have the ability to transfer the heat out of the building.

One of the most important advantages of an effective improvement of insulation level is not only that it reduces the amount of heat losses, but at the same time it makes the indoor environment of a building more thermally comfortable. As a result, the installation of insulation systems can offer remarkable savings in energy bills and reduce the demand for cooling and heating energy [8].

Table 5.3 Types of insulation and current best practice maximum U-values (W/m²K) [9, 10]

Type of insulation		Maximum U-value
Wall insulation [9]	Cavity walls	Depends on cavity width
	Internal wall insulation	0.45
	External wall insulation	0.35
Floor insulation [9]	Suspended timber floors (ST)	0.20–0.25
	Solid underfloor insulation (UF)	
	Solid overfloor insulation (OF)	
Roof insulation [9]	Loft insulation	0.16
	Internal roof insulation	0.16
	External roof insulation	Not specified
	Flat roof insulation	0.25
Window insulation [9]	Replacement windows (RW)	2.0
	Secondary windows (SW)	
	Moveable insulation (MO)	
Door insulation	Glazed doors [9]	2.0
	Half glazed door [10]	3.7
	Insulated half glazed door [10]	1.5
	Solid doors [9]	1.0
	Solid timber door [10]	3.0
	New insulated solid door [10]	1.0

Furthermore, insulation systems can limit the environmental impact of a building over its lifetime [9]. Additionally, insulation systems have the ability to reduce the environmental impact of insulation materials [10].

The basic types of insulation systems are: wall insulation, floor insulation, roof insulation, window insulation and door insulation. Table 5.3 presents the most important subcategories of insulation systems and the maximum U-values for each one.

5.2.2 Glazing

Firstly, the basic role of glazing is that it can provide an outdoor view and it can permit the entrance of daylight in the indoor environment of a building. In addition, glazing can improve the rate and the quality of ventilation [11].

The use of glazing has the ability to eliminate the amount of heat losses in buildings. In particular, when there are high amounts of heat losses in the indoor environment of a building, this means that there are increased demands for energy consumption. As a result, occupants should pay hefty energy bills to cover their

Table 5.4 Characteristics and U-values for different types of glazing [12]

Type of glazing	Characteristics	U-value (W/m^{2}K)
Single glazing	Timber and PVC-U window frames	4.7
	Thermally broken aluminium frames	5.3
	Steel or aluminium frames	5.8
Double glazing (12 mm air filled cavity)	Timber and PVC-U window frames	2.8
	Metal frames	3.4
Double glazing with soft coated low-e glass (12 mm air filled cavity)	Timber and PVC-U window frames	2.1
	Metal frames	2.6
Double glazing with hard coated low-e glass (16 mm air filled cavity)	Timber and PVC-U window frames	2.0
	Metal frames	2.5
Triple glazing with hard low-e coating (16 mm air filled cavity)	Timber and PVC-U window frames	1.6
	Metal frames	2.0

Table 5.5 U-values for half-glazed doors and windows [12]

Half-glazed doors	Characteristics	U-value (W/m^{2}K)
	Timber framed—Single glazing	3.7
	Timber framed—Double glazing	3.0
	Insulated door—Double glazing	1.3
Windows	Characteristics	U-value (W/m^{2}K)
Existing window	Timber window frames	4.7
	Metal window frames	5.8
With secondary glazing	Timber window frames	2.9
	Metal window frames	3.4

needs for heating purposes. Glazing has a key role to control and reduce the heat losses of a structure [11].

The factors that affect the flow rate of heating energy through the glazing of windows are:

- The variation between the temperatures of the two sides of glazing
- The average area that permits the flow
- The insulation properties of the glazing material.

The main types of glazing that are available on market are shown at Table 5.4, as well as the respective U-values. The most common application of glazing in dwellings is its use in windows, roof windows and glazed doors. Table 5.5 presents the U-values for half-glazed doors and windows for different characteristics and glazing types.

5.2.3   Ventilation

Ventilation is responsible for the air removal of indoor environment and then its replacement with the fresh air of outdoor environment. The process of ventilation occurs through openings, cracks and gaps in the envelope of the structure [13]. The main advantage of ventilation is the fact that it can achieve a thermally comfortable environment inside the building [14].

The use of ventilation systems can improve the air quality and the sense of thermal comfort in dwellings. Ventilation systems can optimize the conditions that secure well-being and human productivity [15]. Moreover, they can eliminate the phenomenon of overheating during summer [13] and the domestic condensation [16]. It is also important that the excessive use of ventilation should be avoided, because it can lead to significant amounts of heat losses. As a result, this phenomenon would cause an increase in the amounts of energy bills, fuel consumption and CO_2 emissions [17].

The most important types of ventilation are:

- Natural Ventilation

The natural ventilation occurs through openings and without the installation of ventilation technologies. Wind pressure and air density determine the air quality.

- Mechanical Ventilation

The mechanical ventilation is based on the operation of mechanical systems. In particular, a mechanical system consists of a fan and distribution ductwork. The type of mechanical ventilation can be easily controlled, but it is more complex compared to natural ventilation [18].

- Hybrid Ventilation

The hybrid ventilation combines the natural ventilation and the mechanical ventilation [19].

- Infiltration

Infiltration is a process of air leakage and it is a type of accidental ventilation [20].

The most important advantage of natural ventilation is the fact that when it is used for heating applications, it can achieve higher temperatures in the indoor environment of a building than the air conditioning systems. As a result, natural ventilation may reduce the use of air conditioning systems, which require increased amounts of energy consumption.

On the other hand, mechanical ventilation systems can significantly reduce the amount of energy consumption in dwellings, particularly when they are used for cool- ing applications. Generally, the use of ventilation systems can reduce the amount of heat gains during summer and respectively the amount of heat losses during winter, due to the materials that are used in ventilation systems.

5.2.4 Micro-distributed Technologies

Heat pumps. A heat pump uses renewable heat sources in order to offer heating and cooling loads to buildings [21]. Heat pump systems are technologies which intend to cover the domestic demands for heating or cooling. The operation of a heat pump is based on the fact that heating is transferred from a warmer environment to a cooler environment.

The heating, which is derived from the use of heat pumps, is one of the most efficient solutions in mild and moderate climates. These systems offer a significantly greater heating output than the respective amount of the energy, which is consumed for their operation. The low carbon output of heat pumps is lower than the half of the respective output that is derived from electric, oil and gas heat production, which means that the operation of a heat pump is an environmental friendly energy technology [21].

The two basic categories of heat pumps are: the air source heat pump (ASHP), which uses the latent heat of the air, and the ground source heat pump (GSHP), which uses the latent heat that is stored to the ground [22]. Table 5.6 presents the average cost for a typical air source heat pump (ASHP) and for a ground source heat pump (GSHP).

Solar Photovoltaic. Photovoltaic (PV) can convert the solar energy that they receive into electricity. The basic factor, which determines the energy efficiency of PV panels, is the type of collected solar radiation (direct or diffuse light).

Even though common PVs can last approximately 10–20 years, some PVs, which are available in the market, have a lifetime of 25 years or more. Two significant advantages of PVs are that they do not include moving parts and that their maintenance needs are not demanding [24]. Generally, the efficiency of PVs ranges from 5 to 40%, while the efficiency of a typical PV system ranges from 10 to 15%. Further advantages of PVs, is the fact that the use of a PV system is safe and it does not emit noises [11].

The wide installation of PV systems is an effective way to conserve amounts of natural resources and also to reduce the energy needs [25]. According to studies, the environmental impact of a PV system is lower than the environmental impact of other types of renewable and non-renewable energy technologies, which produce electricity [11].

The basic factor, which determines the energy efficiency of PV panels, is the type of collected solar radiation. The light, which is collected from PV panels, can

Table 5.6 Cost of ASHP and GSHP [23]

Air source heat pump (ASHP) and ground source heat pump (GSHP)	
Type of technology	Cost
Typical system of ASHP	£6000–£10,000
Typical system of GSHP	£9000–£17,000

be direct or diffuse. During a clear day, the percentage of diffuse light is estimated approximately 5% of the total solar radiation. On the other hand, during an overcast day the percentage of diffuse light compared to the total is higher than 5%. In addition, the efficiency of a PV panel depends on their orientation, which means that PV systems should be properly positioned in order to receive the highest potential amount of solar radiation and to achieve their optimal energy efficiency and performance. Ideally, the most efficient orientation of PV panels for achieving the optimal energy performance is between south-east and south-west at an elevation of 30°–40° [24]. Table 5.7 shows the cost, the potential CO_2 reduction and the annual savings in energy bills for a 4 kWp PV system, which is the average domestic solar PV.

Solar thermal. Approximately 7% of total delivered energy in UK, corresponds to the demand of water heating for domestic uses. In addition, the demand of water heating per day for a representative household in UK is 15 kW of delivered energy. Approximately the half of the existing households in UK could have installed a system of solar water heater.

It is estimated that according to this hypothetical plan, almost 12 million of solar thermal technologies or 50 million m^2 of collectors would be required [11]. The basic categories of solar thermal collectors, which are used for domestic applications, are: the flat plate and the evacuated tube collector [22]. Table 5.8 shows the average cost, the potential CO_2 reduction and the annual savings in energy bills for a typical solar thermal system.

Micro CHP. A micro CHP (combined heat and power) is a technology, which can produce heat, hot water and electricity simultaneously [26]. It seems that, a micro CHP system generates increased amounts of heat and power, to the extent that, it has the ability to cover the total amount of electrical and heating needs of a

Table 5.7 Cost and potential CO_2 reduction of PV system [23]

PV system	
Type of technology	A 4kWp PV system
Cost	£5000–£8000
Reduction of CO_2 emissions (tonnes CO_2/year)	1.8 tonnes per year
Annual savings (£/year)	£750–£770 per year

Table 5.8 Cost, potential CO_2 reduction and annual savings of solar thermal system [23]

Solar thermal			
Type of technology	Cost	Reduction of CO_2 emissions (kg CO_2/year)	Annual savings (£/year)
Typical Solar Thermal system	£3000–£5000	For replacing gas: 275 kg CO_2 emissions per year	£60 per year
		For replacing electric immersion heating: 490 kg CO_2 emissions per year	£70 per year

Table 5.9 Efficiency for different types of Micro CHP system [27]

Micro CHP	
Types of micro CHP	Efficiency (%)
Internal combustion engines	25–45
Stirling engines	12–30%
Rankine cycle engines	
• Small engines • Large power stations	• 10 • 40

house. At the same time, its use could effectively replace the installation of a domestic convectional boiler [22].

According to Table 5.9, the most common categories of micro CHP systems are:

- Internal combustion engines

The average lifetime period of large internal combustion engines is greater than 20 years. The lifetime period of smaller internal combustion engines is usually shorter.

- Sterling engines

In most cases, a Sterling engine has small size and it ranges between 1 and 25 kW. There are also larger Sterling engines of 500 kW. The Sterling engines are more ex- pensive compared to other types of micro-CHP systems.

- Rankine cycle engines

A Rankine cycle engine is more cost effective than other prime mover technologies and this is a reason why it is considered as a very competitive prime mover technology. The efficiency of a Rankine cycle engine varies according to the size of engine.

Table 5.9 presents the efficiency of engines for each micro CHP type. The basic categories of micro CHP technologies for domestic purposes are: the internal combustion and the Stirling engines.

Wind Turbine. During the last years, the use of wind turbines is considered as one of the most rapidly growing renewable energy systems worldwide [11]. Wind turbines have the ability to generate amounts of electricity, but without emitting greenhouse gas emissions [28]. There are two basic categories of wind turbines: micro generation and small-scale generation [22]. The domestic wind turbines are small-scale versions of large wind turbines [26]. UK legislation, companies and grant schemes make efforts to promote and encourage the use of domestic wind turbines.

The potential energy generation, which is derived from micro generation wind turbines, is at least 1.5 kW of electricity. The respective value for small-scale wind turbines is 1.5–15 kW of electricity [22]. Moreover, increased amounts of the energy output from wind turbines, which are not used in an instantaneous way in buildings, can be exported [29]. The most common micro wind turbines include small blades of 1.7 m [26].

Table 5.10 Cost and potential CO_2 reduction of domestic wind turbine [23]

Wind turbines	
Type of technology	Cost
Roof-mounted 1 kW microwind system	£3000
2.5 pole-mounted system	£9900–£19,000
6 kW pole-mounted system	£21,000–£30,000
Reduction of CO_2 emissions from a 6 kW system (tonnes CO_2/year)	5.2 tonnes per year

It is estimated that micro wind turbines account almost 17% of the grant aid that intends to support the use of micro generation technologies in the domestic sector [29]. The installation of micro wind turbines is one the most effective ways to promote the global effort in order to reduce the greenhouse gas emissions [30]. Table 5.10 represents the typical cost for three different types of wind turbines and the reduction of CO_2 savings for a 6 kW system of wind turbine.

5.3 Proposed Packages Through Green Deal

Five different dwelling types are examined in order to establish the appropriate recommended Green Deal packages: a semi-detached house, a terraced house, a flat, a detached house and a bungalow. The selected dwelling types are representative of the UK building stock.

The different types of energy saving measures, which are combined in order to examine the potential Green Deal packages, are shown at Table 5.11.

The recommended packages contain the maximum number of measures, which could be combined and they are evaluated in terms of the estimated amount of CO_2 reduction. Each recommended package must satisfy the restriction for the maximum cost of Green Deal package (£10,000). After various computational combinations, three recommendations were selected for each dwelling type. The recommended Green Deal packages could offer the highest potential amount of CO_2 savings. Tables 5.12 and 5.13 include the calculation of total cost and CO_2

Table 5.11 Technologies of proposed Green Deal packages

Types of measures of Green Deal packages			
Type of measure	Insulation	Glazing	Renewable energy technologies
Potential energy saving measures	• Internal • External • Cavity • Floor • Roof/loft • Doors	• Double glazing	• Heat pumps (ASHP or GSHP) • Solar PV • Solar thermal • Micro CHP • Wind turbine

Table 5.12 Proposed packages for semi-detached houses, terraced houses and flats

Dwelling type	Recommended packages		Total cost (£)	CO_2 savings (tC/year)
	No.	Energy saving Measures		
Semi-detached house	1	Cavity insulation Floor insulation Roof/loft insulation Doors insulation Micro chp	9395	3.83
	2	External insulation Cavity insulation Wind turbine	8278	3.35
	3	Cavity insulation Micro CHP	7653	3.23
Terraced house	1	Cavity insulation Floor insulation Roof/loft insulation Doors insulation Micro chp	9011	3.12
	2	Cavity insulation Floor insulation Roof/loft insulation Micro CHP	8591	2.99
	3	External insulation Cavity insulation Floor insulation Roof/loft insulation Doors insulation Wind turbine	9058	2.35
Flat	1	Cavity insulation Floor insulation Roof/loft insulation Doors insulation Micro CHP	8823	3.61
	2	External insulation Cavity insulation Floor insulation Roof/loft insulation Doors insulation Wind turbine	7831	3.29
	3	Internal insulation Cavity insulation Floor insulation Roof/loft insulation Doors insulation Wind turbine	6998	3.25

savings, which is derived from the implementation of these packages. The amount of total cost is given by the sum of individual costs of measures for each package, while the total CO_2 savings are given by the sum of the potential CO_2 reductions.

Table 5.13 Proposed packages for detached houses and bungalows

Dwelling type	Recommended packages		Total cost (£)	CO$_2$ savings (tC/year)
	No.	Energy saving measures		
Detached house	1	Internal insulation Cavity insulation Floor insulation Roof/loft insulation Doors insulation Wind turbine	9755	5.44
	2	Internal insulation Cavity insulation Floor insulation Roof/loft insulation Wind turbine	9230	5.33
	3	Internal insulation Cavity insulation Floor insulation Doors insulation Wind turbine	9011	5.26
Bungalow	1	Cavity insulation Floor insulation Roof/loft insulation Doors insulation Micro CHP	8895	3.77
	2	External insulation Cavity insulation Floor insulation Roof/loft insulation Doors insulation Wind turbine	8249	3.56
	3	Internal insulation Cavity insulation Floor insulation Roof/oft insulation Doors insulation Wind turbine	7334	3.51

5.4 Results

Table 5.14 provides an overview of the general profile of housing stock in Britain in 2011. The profile of housing stock presents the number of dwellings and the number of household occupants and also the average and total floor area for the dwelling types that are examined in this study. In addition, Table 5.14 shows the percentage of housing stock for each dwelling type. The number of dwellings and the percentage of housing stock are presented according to the English Housing Survey for 2011–2012. In 2011, it is estimated that the total amount of dwellings in Great Britain was 22,754,000 [31].

Table 5.14 Existing housing stock in Britain in 2011

	Number of dwellings	% of housing stock	Household occupants	Average floor area (m²)	Total floor area (m²)
Semi detached	5917	26.0	2	89	526,613
Terrace	6428	28.2	2–3	79	507,812
Flat	4628	20.4	2	61	282,308
Detached	3786	16.6	3	104	393,744
Bungalow	1996	8.8	2	67	133,732

Table 5.15 Standards of U-values for existing and refurbished dwellings

Existing dwellings—heat losses/U-values (W/m²K)					
	Ground floor	Walls	Roofs	Windows	External doors
Solid wall	0.70	2.10	1.00	Single 4.80	3.00
Cavity wall	0.70	1.50	0.40	½ double, 6 mm 3.10	3.00
Refurbished dwellings—heat losses/U-values (W/m²K)					
Refurbished	0.25	0.35	0.16	1.30	1.00
AD Part L1B	0.25	0.35	0.25	2.20	2.20

Table 5.15 presents the standards of U-values for existing and refurbished dwellings. Firstly, in terms of the existing houses, Table 5.15 includes the U-values for solid and cavity wall. Secondly, the U-values are presented for an average refurbished house and according to AD Part L1B. The U-values, which are shown at Table 5.15, are the desired values after the implementation of a Green Deal Package.

5.4.1 Economic Analysis

Figure 5.1 shows the capital cost in pounds (£) for different energy measures of the five types of houses and Fig. 5.2 presents the capital cost of renewable energy technologies. It is assumed that the renewable energy technologies do not depend on the dimensions of each dwelling type. The highest capital cost of various insulation types and glazing can be noticed at detached dwellings, because the cost of these measures depends on the average floor area of the building. For this reason, the flat has the least cost for the installation of glazing and insulation (Fig. 5.1).

The installation of a GSHP or a solar PV system is remarkably expensive (Fig. 5.2) and it approaches the maximum permitted cost of Green Deal (£10,000). As a result, the packages that include these measures should be combined with low-cost measures in order to not exceed the upper limit cost of Green Deal. For all the dwell- ing types except from flats, glazing has an increased capital cost, which is

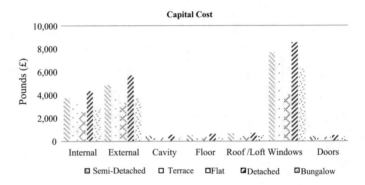

Fig. 5.1 Capital cost of insulation and glazing for different dwelling types

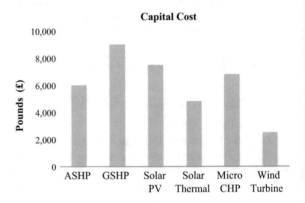

Fig. 5.2 Capital Cost of renewable energy systems

higher than £6000. The expensive capital cost of glazing is the reason why the proposed Green Deal packages do not contain the installation of glazing. In case of internal or external insulation, the installation cost is high compared to other energy saving measures. The size of external wall areas for a particular building, deter- mines the cost of these insulation types.

The capital cost for the majority of renewable energy technologies and windows insulation systems is significantly high compared to insulation. In particular, ground source heat pump is the most expensive measure for all the types of dwellings (£9000). On the other hand, the measures with the least installation cost are cavity, floor and roof/loft insulation. Even if the aforementioned measures cannot offer significant reduction of CO_2 emissions, the advantage of their low cost makes their potential use more flexible. As a result, these energy saving measures are consid- ered as advantageous measures for the selection of Green Deal packages.

5.4.2   Carbon Savings

Figure 5.3 presents the amounts of CO_2 savings, which are derived from the use of insulation and glazing for five different dwelling types, while Fig. 5.4 shows the potential CO_2 savings from the use of renewable energy technologies.

According to Fig. 5.3, the installation of external and internal insulation in detached houses offers the highest amounts of carbon savings (the respective CO_2 reductions are 2.36 and 2.26 tC per year). The annual CO_2 reduction due to the use of micro CHP is also remarkable (2 tC per year) (Fig. 5.4). In addition, according to Fig. 5.3, the use of cavity insulation in detached dwellings could offer high annual carbon savings, which can climb to 1.9 tonnes. In semi-detached houses (Fig. 5.3), the installation of cavity, internal or external insulation can offer high amounts of carbon savings. As a result, for semi-detached houses, the installation of cavity, internal or external insulation is advantageous and recommended for the selection of the appropriate Green Deal package.

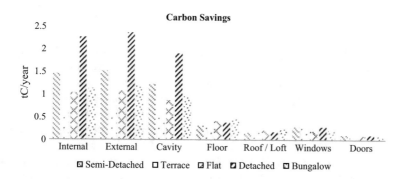

Fig. 5.3 CO_2 savings from the use of insulation and glazing for different dwelling types

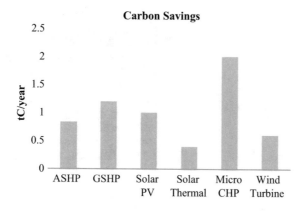

Fig. 5.4 CO_2 savings from renewable energy systems

The use of roof/loft, doors and windows insulation cannot be considered as priority measures (for each dwelling type), due to the low amount of CO_2 savings that they offer, but they are suitable to be used as additional and cost saving measures.

5.4.3 Cost and CO_2 Savings of Recommended Green Deal Packages

Figure 5.5 presents the total cost of the three recommended Green Deal packages for each dwelling type. The final costs of proposed packages do not significantly differ between each other. As it has already been mentioned, the final packages combine the maximum number of measures for each dwelling type and as a result, the costs of proposed packages approach the maximum permitted cost of Green Deal (£10,000).

Table 5.16 shows the packages, which offer the maximum and the lowest amount of CO_2 savings. Even if flats have less average floor area, their CO_2 savings are higher than terraced dwellings after the implementation of these recommendations.

The recommended packages of Fig. 5.5 are considered as recommended, because their implementation could offer the maximum reduction in the amount of carbon emissions. For the majority of dwelling types, the first recommended package is the most expensive. Generally, the most expensive package is the first recommendation for detached dwellings (£9755). The minimum cost between all the proposed packages belongs to the third recommended package of flats (£6998).

Small dwellings, such as flats, can achieve significant carbon savings with the selection of a low cost Green Deal package. The packages for dwellings with higher average floor area (for example detached houses) have a remarkably higher cost, as the installation of insulation and glazing technologies is expensive.

Fig. 5.5 Cost of recommended packages

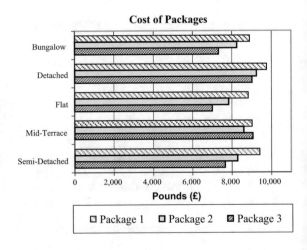

Table 5.16 Recommended packages that offer the maximum and the lowest amount of CO_2 savings

	Maximum CO_2 savings	Lowest CO_2 savings
Dwelling type	Detached dwellings	Terraced dwellings
No of recommendation	1st recommendation	3rd recommendation
Energy saving measures of package	• Internal insulation • Cavity insulation • Floor insulation • Roof/loft insulation • Doors insulation • Wind turbine	• External insulation • Cavity insulation • Floor insulation • Roof/loft insulation • Doors insulation • Wind turbine
Total CO_2 savings	5.44 tC/year	2.35 tC/year

Fig. 5.6 CO_2 savings of recommended packages

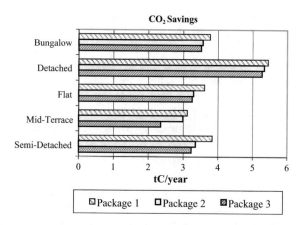

The package that offers the maximum amount of CO_2 savings (5.44 tC/year) can be found at detached dwellings and it includes: internal insulation, cavity insulation, floor insulation, roof/loft insulation, doors insulation and wind turbine (Table 5.16). On the other hand, the package that offer the lowest amount of CO_2 savings (2.35 tC/year) can be found at terraced dwellings and it includes: external insulation, cavity insulation, floor insulation, roof/loft insulation, doors insulation and wind turbine (Table 5.16).

Figure 5.6 compares the amount in the reduction of CO_2 savings (in tC/year) after the implementation of the three recommended Green Deal packages for each dwelling type. According to the proposed Green Deal packages of Fig. 5.6, even if a typical flat is smaller than a typical terraced dwelling, the amount of carbon savings of a flat is higher than the respective value for terraced dwellings. The proposed Green Deal packages for large houses (for example detached dwellings) have also increased total cost. Moreover, the particular types of dwellings could offer the highest achievable amount of carbon savings.

5.5   Discussions and Conclusions

In order to reduce the energy consumption and protect the environment from further climate change, governments promote policies that encourage the installation of renewable energy technologies. Green Deal is a UK environmental policy, which has been introduced to the UK in 2012 and it provides support to the installation of various energy saving technologies [32]. This study has examined the potential Green Deal packages for five different types of dwellings and has also evaluated the packages in terms of their total cost and the potential CO_2 reduction that they can offer.

Generally, the selection of the recommended Green Deal packages, their total cost and the amount of CO_2 savings that they can achieve, depend on the average floor area of each house and the dwelling type. The most important restriction for the implementation of a Green Deal package is the fact that £10,000 is the maximum cost of a package; a limit, which is determined from the UK government. As a result of this restriction, small dwellings (for example flats or bungalows) have a variety of potential packages to choose in order to cover their needs. In particular, for flats and bungalows, there are various combinations of energy savings measures, which have less cost than the upper limit cost of Green Deal. On the contrary, dwellings with higher average floor area (for example detached and semi-detached houses) have less number of available options for the selection of a package. This can be explained due to the high cost of insulation and glazing for large buildings in comparison to the respective cost for small buildings.

The selection of the appropriate Green Deal package is unique and it depends on the energy needs and the preexisting situation of the dwelling. The composition of the recommended packages is varied according to the energy conditions and the total area of the building. In addition, the restriction for the upper limit cost of each package is a crucial factor that affects the number of the recommended Green Deal packages. For example, the number of recommended packages is limited for an old large house with high-energy losses and/or lack of insulation and/or ventilation technologies, due to the maximum limit cost of the potential package. Furthermore, the number of potential Green Deal packages is higher for a small house with low heat losses. Another aspect for the effective application of a Green Deal package is the assiduous examination of the required conditions for the installation of renewable energy systems (for example the examination of orientation, location, etc.). A Green Deal advisor evaluates the aforementioned factors and he determines the final recommended package.

The installation of insulation is a remarkably effective measure of Green Deal. Insulation is easily combined with a plenty of measures (for example glazing or renewable energy technologies), due to its relatively low cost. Each recommended package contains only one type of renewable energy technologies (e.g. a wind turbine or a micro CHP) due to their increased cost. The use of a GSHP cannot be easily combined with other measures, because its installation is the most expensive. It can be seen that the installation of ASHP, solar PV or solar thermal requires lower cost compared to GSHP, but at the same time the potential carbon reduction that they offer is also lower than GSHP.

References

1. Charles, Ph., Symons, K., Wardle, J.: Rethinking Housing Refurbishment, Construction & Refurbishment 2008: Building a Future, MBE KTN Masterclass (2008)
2. Pan, W., Garmston, H.: Building regulations in energy efficiency: compliance in England and Wales. Energy Policy **45**, 594–605 (2012)
3. Pettifor, H., Wilson, C., Chryssochoidis, G.: The appeal of the green deal: Empirical evidence for the influence of energy efficiency policy on renovating homeowners. Energy Policy **79** (2015), 161–176 (2015)
4. Guertler, P.: Can the green deal be fair too? Exploring new possibilities for alleviating fuel poverty. Energy Policy **49**(2012), 91–97 (2012)
5. Dowson, M., Poole, A.D., Harrison, D., Susman, G.: Domestic UK retrofit challenge: barriers, incentives and current performance leading into the Green Deal. Energy Policy **50**(2012), 294–305 (2012)
6. DECC: Department of Energy and Climate Change: What measures does the Green Deal cover? (2011)
7. DECC: Department of Energy and Climate Change: The Green Deal—a summary of the Government's proposals (2010)
8. BRUMFA: Energy Efficient Buildings Retrofit Solutions (2003)
9. Energy Saving Trust: Energy Efficiency Best Practice in Housing: Advanced insulation in housing refurbishment (2005)
10. Energy Saving Trust: Energy Efficiency Best Practice in Housing: Refurbishing Timber-Framed Dwellings (CE59) (2004)
11. Boyle G.: Renewable Energy: Power for a Sustainable Future, Second edition. Oxford University Press, Oxford (2004)
12. Energy Saving Trust: Refurbishment site guidance for solid-walled houses—windows and doors, Good Practice Guide 295 (2002)
13. Energy Saving Trust: Energy efficient ventilation in dwellings—A guide for specifiers, Good Practice Guide 268: CE124 (2006)
14. Qingyan, C.: Ventilation performance prediction for buildings: a method overview and recent applications. Build. Environ. **44**, 848–858 (2009)
15. Dimitroupoulou, C.: Ventilation in European dwellings: a review. Build. Environ. **47**(2012), 109–125 (2012)
16. Energy Saving Trust: Energy Efficiency Best Practice in Housing: Refurbishing cavity-walled dwellings, (CE57) (2004)
17. Energy Saving Trust: Energy Efficiency Best Practice in Housing: Energy efficient loft extensions (CE120) (2005)
18. Department of Health: Heating and Ventilation Systems: Health Technical Memorandum 03-01: Specialized ventilation for healthcare premises. Des validation, Part A (2007)
19. Csaba, Szikra: Hybrid Ventilation Systems. Budapest University of Technology and Economics, Faculty of Architecture, Hungary (2004)
20. NRCS: Natural Resources Conservation Service: Soil Quality Indicators: Infiltration (1998)
21. Abdeen, M.O.: Ground-source heat pumps systems and applications. Renew. Sustain. Energy Rev. **12**(2008), 344–371 (2006)
22. Kitcher C: A Practical Guide to Renewable Energy—Microgeneration Systems and their installation (2012)
23. Energy Saving Trust Website: http://www.energysavingtrust.org.uk/

24. Jardine, C.N., Lane, K: Photovoltaics in the UK: An Introductory Guide for New Consumers, Environmental Change Institute, Oxford. Available at: www.eci.ox.ac.uk/research/energy/downloads/pv-inthe-uk.pdf (2003)
25. Razykov, T.M., Ferekides, C.S., Morel, D., Stefanakos, E., Ullal, H.S., Upadhyaya, H.M.: Solar Photovoltaic Electricity: Current Status and Future Prospects. Sol. Energy **85**(2011), 1580–1608 (2011)
26. Green Deal Initiative: Available at: http://www.greendealinitiative.co.uk/ (2013)
27. U.S. Department of Energy: Micro-CHP Systems for Residential Applications—Final Report, Prepared by United Technologies Research Center (2006)
28. James, P.A.B., Sissons, M.F., Bradford, J., Myers, L.E., Bahaj, A.S., Anwar, A., Green, S.: Implications of the UK field trial of building mounted horizontal axis micro-wind turbines. Energy Policy **38**(2010), 6130–6144 (2010)
29. Peacock, A.D., Jenkins, D., Ahadzi, M., Berry, A., Turan, S.: Micro wind turbines in the UK domestic sector. Energy Build. **40**(2008), 1324–1333 (2008)
30. Zhe, Li, Fergal, Boyle: Reynolds Anthony, Domestic application of micro wind turbines in Ireland: investigation of their economic viability. Renew. Energy **41**(2012), 64–74 (2012)
31. Department for Communities & Local Government: English Housing Survey, Headline Report 2011–2012 (2013)
32. Tr, Sweetnam, Spataru, C., Cliffen, B., Sp, Zikos, Barett, M.: PV System Performance and the Potential Impact of the Green Deal Policy on Market Growth in London, UK. Energy Procedia **42**(2013), 347–356 (2013)

Chapter 6
Analysing the Contribution of Internal Heat Gains When Evaluating the Thermal Performance of Buildings

Rainer Elsland

Abstract To evaluate the useful energy demand for space heating purposes, norm-based bottom-up models are applied that capture the energy-related characteristics of buildings. As the electricity demand per household attributed to appliances and lighting has increased substantially in the EU27 over the last two decades, the question arises whether the static, norm-based approach is underestimating the contribution of internal heat gains to covering useful energy demand. To analyse the impact of dynamic internal heat gains in the residential sector, a bottom-up model is applied that covers the EU27 building stock with a country-specific typology. The study reveals that the norm-based approach underestimates the contribution of internal heat gains to covering thermal heat demand. Comparing the countries throughout the EU27 climate zones indicates that, on average, the static and the dynamic share of internal heat gains up to 2050 vary in a range from 20 to 70%.

Keywords Bottom-up approach · Residential sector · Buildings · Internal heat gains · Scenario analysis

6.1 Background and Objectives

Lowering energy demand is known to be the most significant driver in the attempt to mitigate climate change and meet the energy and climate policy targets of the European Union [1, 2]. In this context it is of great interest that space heating demand in the residential sector is set to significantly decline over the coming decades. In 2010, 8595 PJ were attributed to residential space heating in the EU 27, which is equivalent to 17.8% of the total final energy demand (48,291 PJ) [3]. Against this background, improving the energy performance of buildings can contribute substantially to reducing energy demand. European and national norms

R. Elsland (✉)
Fraunhofer Institute for Systems and Innovation Research ISI, Karlsruhe, Germany
e-mail: Rainer.Elsland@isi.fraunhofer.de

© Springer International Publishing AG 2017   139
J. Littlewood et al. (eds.), *Smart Energy Control Systems for Sustainable Buildings*,
Smart Innovation, Systems and Technologies 67,
DOI 10.1007/978-3-319-52076-6_6

are applied to evaluate the useful energy demand for space heating purposes of a single building, e.g. DIN V 4108-6, DIN V 4701-10 and DIN EN ISO 13790, that describe the energy-related characteristics of a building based on physical and behavioural parameters [4–6]. When analyzing buildings' energy demand from an energy economics perspective, bottom-up vintage models are applied [7–10]. These bottom-up models capture the building typology of entire countries by construction period, building type and building standard, where each segment is represented by a reference building.

However, although norm-based modelling is a well established approach in bottom-up analysis, this methodology displays a crucial limitation in terms of its negligence of the dynamic characteristics of internal heat gains. An analysis of the development of residential appliances and lighting over the last two decades reveals that their electricity demand per household has increased substantially in each EU member state by around 50% [3]. In spite of the fact that energy policy instruments like the Ecodesign Directive have been implemented during the last decade in an attempt to reduce energy use by setting minimum efficiency standards at the design phase, increased ownership of appliances has still led to rising electricity demand [11]. Against this background, applying a static approach to internal heat gains could lead to a systematic underestimation of their contribution to covering useful energy demand. This methodological limitation becomes even more significant in nearly Zero-Energy Buildings (nZEB), where the relative share of internal heat gains is even larger.

To analyse the impact of the dynamic development of internal heat gains on the useful energy demand in the residential sector of the EU27, the bottom-up vintage model FORECAST-Residential is applied, which covers the European building stock by country. In the framework of this study, internal heat gains are analysed in detail distinguished by the heat given off by different types of appliances, lighting and residents. To the knowledge of the authors, no prior study has focused on this research question on such a broad regional scope. We begin by discussing the methodological concept of the bottom-up approach (Sect. 6.2). In Sect. 6.3 a case study is conducted for the residential building stock of the EU27 on a country basis until 2050. The case study consists of two explorative scenarios; the reference scenario represents the norm-based modelling of internal heat gains and the second scenario considers the dynamic development of appliances, lighting as well as residents. The results are then discussed and conclusions drawn (Sect. 6.4).

6.2   Methodological Approach

6.2.1   Structural Framework

For the analysis the simulation based bottom-up model FORECAST-Residential is applied, which models the useful energy demand for space heating purposes of the

EU27[1] by country up to 2050. FORECAST-Residential is designed as a vintage stock model, which enables a detailed modelling of the stock turnover taking into account regulatory requirements. In this study the country-specific useful energy demand is calculated differentiated by construction period (<1960, 1961–1990, 1991–2008, 2009–2020, 2021–2050), which are in turn divided into building types [single-family-houses (SFH), multi-family-houses (MFH)] and thermal efficiency standards (5 varying standards). In terms of thermal efficiency, building segment packages are defined that represent low, medium or high thermal quality levels for each building element:

- Standard 1: Average thermal performance of a building in 2008 (based on [14–16])
- Standard 2: Replacement of windows (low)
- Standard 3: Replacement of windows (low) and additional refurbishment of the walls (low)
- Standard 4: Replacement of windows (high), refurbishment of the walls (medium) and refurbishment of the roof (medium)
- Standard 5: Replacement of all building elements (high).

Distinguishing the new building stock into buildings constructed before and after 2020 is related to the fact that major policy regulations regarding the energy performance of buildings are defined for the year 2020 (e.g. EPBD recast) [17]. In total, considering the building typology in this way results in 50 reference building segments (see Sect. 6.2.2) per country and thus 1350 building segments for the EU27 (see Fig. 6.1).

An additional module is integrated into the building stock model (see Sect. 6.2.3) to evaluate the dynamic impact of internal heat gains. The internal heat gains module covers large appliances (including: refrigerators, freezers, washing machines, dryers, dishwashers, stoves), ICT (including: desktop-computers, PC-screens, laptops, televisions, set-top-boxes, modems/routers), lighting (including: incandescent lamps, halogen lamps, energy saving lamps, fluorescent lamps, LEDs) as well as 'Others' (this category captures small appliances such as vacuum cleaner and all end-uses that are not explicitly modelled). The internal heat gains module has three hierarchical levels: appliances (e.g. television), technologies (e.g. LCD, plasma) and efficiency classes (e.g. A++, A+) (see Fig. 6.2). The energy demand of the end-uses is captured by techno-economic (e.g. operating power, investments) and user behavioural parameters (e.g. operating hours). The high level of disaggregation makes it possible to also consider rebound effects, policy regulations, saturation effects and investment barriers.

---

[1]FORECAST is a modeling platform that captures the final energy demand of the sectors industry, residential, tertiary, transport and agriculture for the EU 27 + 3 (3: Norway, Switzerland, Turkey) by country up to 2050 [1, 12, 13].

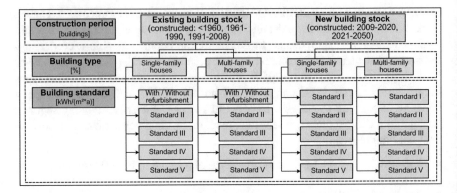

Fig. 6.1 Overview of the structure of the building module

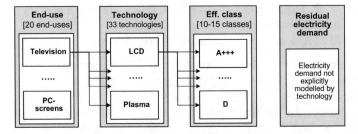

Fig. 6.2 Overview of the structure of the dynamic internal heat gains module

6.2.2 Modelling Procedure

The modelling procedure of the norm-based building module and the dynamic internal heat gains module is illustrated in Fig. 6.3. The socio-economic parameters database provides the framework data for the modelling process. In both modules, the investment decision is based on the Total Cost of Ownership (TCO) and the energy policy framework as well as technological preferences which are considered in terms of diffusion restrictions (e.g. demolishment of buildings at earliest 20 years after construction, replacement of appliances at earliest after 40% of their lifetime). The TCO is calculated by Eq. (6.1):

$$TCO = A(I, DI, dr, re) + S + MC + EC \ [\text{€}] \tag{6.1}$$

where *TCO* is the Total Costs of Ownership, *A* is the annuity, *I* is the investment sum, *DI* is the disposable income, *dr* is the discount rate, *re* is the reinvestment cycle, *S* is the subsidy, *MC* is the maintenance costs and *EC* is the energy costs.

In the norm-based building module, the number of demolished buildings is based on their age distribution and a predetermined demolition rate [18].

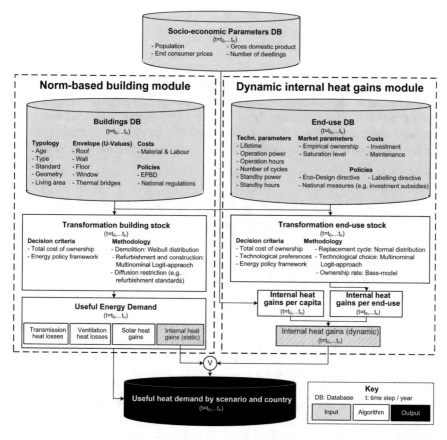

Fig. 6.3 Overview of the modelling procedure of the norm-based building and the dynamic internal heat gains module

The number of new buildings and the refurbishment of existing buildings are calculated based on a multinomial Logit-approach [19] by Eq. (6.2):

$$ms_i = \exp^{-\lambda_i \frac{TCO_i}{\frac{1}{z}\sum_{i=1}^{z} TCO_i}} [\%] \qquad (6.2)$$

where *ms* is the market share of alternative *i*, λ is the heterogeneity of the market, *TCO* is the Total Costs of Ownership and *z* is the number of alternatives.

In general, the dynamic internal heat gains module has a similar structure to the building module apart from the fact that a different set of methodologies is applied: the replacement cycles are captured by a normal distribution function [19], the technological choice is determined using a multinomial Logit-approach [19] and the development of the ownership rate is plotted using a Bass-model [20]. The Bass model is calculated by Eq. (6.3):

$$ow_{i,t} = ow_{i,t-1} + (m_i - ow_{i,t-1}) \cdot \left(a_i + b_i \cdot \frac{ow_{i,t-1}}{m_i} \right) [\%] \qquad (6.3)$$

where *ow* is the ownership rate of alternative *i* in period *t*, *m* is the saturation level, *a* is the innovation coefficient and *b* is the imitation coefficient.

Depending on the scope of the analysis, the internal heat gains are evaluated in a static (norm-based building module) or in a dynamic (dynamic internal heat gains module) manner. In Fig. 6.3 an overview of the modelling procedure of the norm-based building and the dynamic internal heat gains module is presented.

6.2.3 Norm-Based Modelling of Useful Energy Demand for Space Heating Purposes

The useful energy demand per building segment is calculated based on the heating period method taken from the German norms DIN V 4108-6 and DIN V 4701-10 (German Institute for Standardization) [4, 5] which is comparable to the heating and cooling period method of DIN EN ISO 13790 and consequently conform to EU regulations [6]. The annual useful energy demand is the amount of energy supplied to a building during the heating period in order to obtain the target indoor temperature. According to DIN V 4108-6 and DIN V 4701-10, the useful energy demand is calculated by Eq. (6.4):

$$Q_h = (H_T + H_V) \cdot F_{DDF} - \eta_n \cdot (Q_I + Q_S) \, [\text{kWh/a}] \qquad (6.4)$$

where Q_h is the annual useful energy demand of a single building, H_T is the transmission heat loss, H_V is the heat ventilation loss, F_{DDF} is the degree day factor, η_n is the utilization factor including internal and solar heat gains assumed to be 0.95 [4], Q_I is the internal heat gain and Q_S is the solar heat gain. The utilization factor reduces the total gains to the share that covers the thermal heat demand of a building. In terms of internal heat gains the utilization factor considers that end-use power is only partially transferred into convection and radiation and that the location of an end-use is not necessarily located in the heated part of a building (e.g. washing machines are often located in unheated basements).

The transmission heat loss is derived from the heat transfer to the outer surface of the building elements and is therefore a measure of the thermal quality of a building envelope. The thermal quality of an element is defined by its U-value which indicates the level of insulation (W/m²K). Transmission heat losses are calculated based on Eq. (6.5):

$$H_T = \sum_{i=1}^{n} (F_{x,i} \cdot U_i \cdot A_i) + \Delta U_{TB} \cdot A_{tl} \, [\text{W/K}] \qquad (6.5)$$

where H_T is the transmission heat loss, $F_{x,i}$ is the temperature correction factor for each building element i, U_i is the mean U-value of a building element i, A_i is the surface area of each building element i and the additional loss caused by thermal bridging is defined by ΔU_{TB} as the thermal bridge correction factor (0.05 W/m^2 K) [4] and A_{tl} as the heat transmitting surrounding area. The second part of the heat losses is caused by ventilation losses calculated by

$$H_V = c_{air} \cdot n_{air} \cdot V \, [W/K] \tag{6.6}$$

where H_V is the ventilation heat loss, c_{air} is the volume-specific heat capacity of the air, n_{air} is the airflow rate and V is the gross air volume. For SFH and MFH with up to three storeys V is equal to 0.76 and for all other buildings 0.80 [4]. Internal heat gains are caused by the waste heat of electrical end-uses and the heat dissipation of residents, respectively. The calculation of internal heat gains is described in Eq. (6.7):

$$Q_I = 0.024[kh/d] \cdot q \cdot l_{HP} \cdot A_{ref} \, [kWh/a] \tag{6.7}$$

where Q_I is the internal heat gain, q is the average thermal output of internal heat sources, l_{HP} is the length of the heating period and A_{ref} is the energy reference area of the building [4]. The average thermal output of internal heat sources is considered to be constant with 2.5 W/m^2 for SFH and 3.2 W/m^2 for MFH which varies from the norm according to the analysis from IWU [21]. These average values are derived from a field test where a similar ownership of end-uses and resident occupancy is assumed in a SFH and a MFH and thus the difference is the result of the energy reference area ratio per dwelling. Solar heat gains assume complete exposure to sunlight as calculated by

$$Q_S = \sum_{i=1}^{n} (F_i \cdot g_i \cdot A_W \cdot I_{sol}) \, [kWh/a] \tag{6.8}$$

where Q_S is the solar heat gain, F_i is the solar radiation reduction factor for each building element i, g_i is the total energy transmittance of glazing type in case of vertical insolation for each building element i, A_W is the area of windows corresponding to the effective collector area and I_{sol} is the average global irradiation during the heating period depending on orientation. According to DIN V 4108-6 [4] the degree day factor is calculated based on the amount of degree days and a factor of 0.95 for the night setback (time-limited heating) calculated by

$$F_{DDF} = 0.024 \, [kh/d] \cdot (\theta_{i,eff} - \theta_{e,HP}) \cdot l_{HP} \, [kKh/a] \tag{6.9}$$

where F_{DDF} is the degree day factor, $\theta_{i,eff}$ is the effective room temperature during the heating period, $\theta_{e,HP}$ is the average outdoor temperature during the heating period and l_{HP} is the length of the heating period. The effective room temperature depends on the night setback factor (time-limited heating), the space-limited heating factor considering non-heated spaces and an user-specific correction factor.

6.2.4 Modelling Dynamic Internal Heat Gain Development

Analysing the literature shows that in most cases, an average value is taken for internal heat gains either dependent or independent on the type of building [4, 21–25].[2] A more detailed approach is taken by [27, 28], who analysed the drivers of internal heat gains by distinguishing between the heat dissipation of residents and end-uses. Depending on the subject of research, the data basis for internal heat gains results from modelling [24, 27], measurements [23, 29] or surveys [30] and varies widely in a range from 1 to 5 W/m^2. According to [27, 28], the dynamic development of internal heat gains comprises the heat given off by residents, appliances and lighting (Eq. 6.10)

$$Q_{I,t} = Q_{I,e,t} + Q_{I,r,t} \, [\text{kWh/a}] \tag{6.10}$$

where $Q_{I,t}$ is the total internal heat gain, $Q_{I,e,t}$ is the internal heat gain from appliances as well as lighting and $Q_{I,r,t}$ is the internal heat gain from the presence of residents. Thus, Eq. 6.10 replaces Eq. 6.7 when analysing the impact of a dynamic development of internal heat gains. The internal heat gain from appliances as well as lighting is related to the energy reference area of a building and calculated by

$$Q_{I,e,t} = \frac{l_{HP}}{24 \left[\frac{h}{d}\right] \cdot 365 \left[\frac{d}{a}\right]} \cdot \left(\sum_{i=1}^{m} p_{i,t} \cdot d_i \cdot ow_{i,t} \cdot \frac{\eta_{m,i}}{\eta_n} \right) \, [\text{kWh/a}] \tag{6.11}$$

where $Q_{I,e,t}$ is the internal heat gain from appliances as well as lighting, l_{HP} is the length of the heating period, $p_{i,t}$ is the power of an end-use i distinguished by operation and standby, d_i is the running time of an end-use i distinguished by operation and standby per day, $ow_{i,t}$ is the ownership rate of a dwelling by end-use i and $\eta_{m,i}$ is the modified utilization factor for the inclusion of internal heat gains in relation to η_n the norm-based utilization factor [see Eq. (6.4)]. As a result, the utilization factor ratio adjusts the general norm-based default value of 95% to the value of 80% based on the findings of [31–34]. In addition, an end-use specific adjustment to the level of 25% is applied for washing machines and dishwashers, as most of the heat they generate is discharged with the drain water [29]. The amount of thermal heat provided by residents is captured as follows:

$$Q_{I,r,t} = l_{HP} \cdot r_t \cdot d \cdot h \, [\text{kWh/a}] \tag{6.12}$$

where $Q_{I,r,t}$ is the internal heat gain from the presence of residents, l_{HP} is the length of the heating period, r_t is the number of residents per dwelling, d is the average heat dissipation of 70 W per capita of a resident [25, 33, 35] and h is the average daily period spent by a resident in a dwelling, which is assumed to be 12 h here [33].

---

[2]The modelling concept of dynamic internal heat gains is also discussed in [26].

6.3 Case Study

6.3.1 Calibration Basis

To conduct a model-based analysis, a consistent dataset needs to be derived based on recently published studies (e.g. ENTRANZE), commercial databases (e.g. ODYSSEE database) and non-commercial databases (e.g. TABULA database) [3, 14, 15]. The core data of the analysis is the number of dwellings and the living areas distinguished by SFH and MFH. In this context, the living area provided by the ODYSSEE database are assumed to be conditioned living areas. As ENTRANZE also builds upon ODYSSEE data, building stock data are taken from this source [14]. Information related to building physics for reference buildings that represent building standards is based on TABULA including building geometry and U-values for the wall, roof, windows and floor area [15]. To produce a detailed energy economic analysis, a further differentiation of the buildings by construction period is required and is taken from this source [16]. An overview of the dwelling stock and living area data by building type for 2008 is provided in Figs. 6.4 and 6.5. This country-specific data gives an indication of the heterogeneity of the EU27 residential building stock.

The impact of dynamic internal heat gains on the development of useful energy demand is essentially determined by the number of appliances per household (ownership rate). Figure 6.6 shows the average ownership rates of the major appliances in the residential sector for the EU27 in 2008. The highest level of ownership is observed for televisions with 107%, followed by refrigerators with

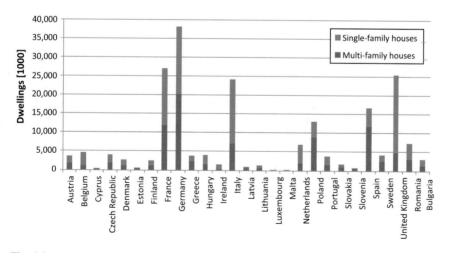

Fig. 6.4 Dwelling stock data by building type in the EU27 in 2008

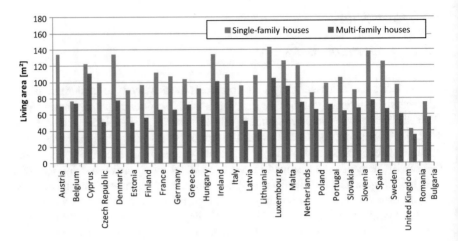

Fig. 6.5 Average living area by building type in the EU27 in 2008

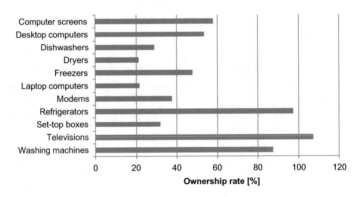

Fig. 6.6 Ownership rates of the major appliances for an average household in the EU 27 in 2008

97% and washing machines with 88%. The other appliances all have an ownership rate below 60% on average in 2008.

6.3.2 Scenario Definition

The explorative scenario analysis examines the useful energy demand of the EU27 residential sector in the period 2008–2050. The reference scenario (REF-S) evaluates the internal heat gains based on the static approach, and the second scenario (DYN-S) analyses the impact of dynamic internal heat gains. To ensure a high level of comparability, the scenarios only differ in terms of the approach used to evaluate

Table 6.1 Overview of the socio-economic framework parameters in the EU27 for 2008–2050 [1, 36]

	Unit	2008	2020	2030	2040	2050
Population	million	495	514	520	522	516
Dwellings	million	203	220	229	235	238
Living area	million m^2	18,836	20,911	22,320	23,571	24,538
Gross domestic product (GDP)	Bn. € (2005)	11,256	14,163	16,824	19,400	22,462
Persons per dwelling	Pers./HH	2.44	2.34	2.27	2.22	2.17
Living area per person	m^2/Person	38.05	40.68	42.92	45.16	47.55
GDP per person	€/Person	22,739	27,554	32,354	37,165	43,531

the internal heat gains. The socio-economic framework parameters for the scenario analysis depicted in Table 6.1 are taken from a study conducted by the Energy System Analysis Agency (ESA^2) for the EU27 by country [1, 36].

6.3.3 Energy Policy Framework

When analysing long-term scenarios, the underlying energy policy framework essentially influences the results. Empirical analysis shows that especially regulative measures in the form of minimum energy performance standards have a significant influence on the development of energy demand.

In relation to buildings, these minimum standards are set by the Energy performance of buildings Directive [37] that obliges all EU Member States are obliged minimum energy performance standards for existing buildings in the case of major renovations and for new constructions [38]. Furthermore, minimum requirements need to be satisfied for each building element (roof, wall, floor, window), although these vary among the EU countries. The highest requirements for building elements can be found in Germany, Austria and Denmark (see Table 6.2 in the Appendix). In addition, national subsidy programmes are considered in this analysis (see Table 6.3 in the Appendix).

Relating to the development of internal heat gains, the Ecodesign Directive set the legal framework for defining of minimum energy performance standards of energy-related products for the European domestic market [11]. Before a product is included in the Ecodesign Directive, a preliminary study is done with a market and a technical analysis of all the product groups covered by the directive. As of March 2013, more than 20 implementing regulations had become law and several other products were in the process of regulation. Furthermore, the Labelling Directive is considered. An energy label is an instrument that informs consumers and thus helps consumers choosing products which save energy and thus money. As of March 2013, delegated regulations for 7 products have been published. An overview of the ecodesign and labelling regulations relevant for the residential sector (heating-related requirements excluded) is illustrated in Table 6.4 in the Appendix.

6.3.4   Results

With reference to large appliances, the ownership rate only increases to a minor extent until 2050, except for dishwashers and dryers (Fig. 6.7). Especially for refrigerators and washing machines, mainly replacement investments will occur in the upcoming decades. The average saturation level of large appliances lies in a range from 50 to 110%. In contrast, the ownership rate of ICT appliances is assumed to rise significantly for all the observed appliances. The highest dynamics can be witnessed for desktop computers, set-top boxes and modems. The highest ownership rate of ICT appliances can be seen for televisions, which reaches 152% in 2050. The saturation level for ICT devices scatters in a range from 70 to 160%. The number of lighting points per dwelling ranges from 15 to 30. In analogy to large appliances and ICT, the exact value of the saturation level depends on the type of end-use and the country.

Analysing the specific energy demand development shows that the efficiency improvement ranges between 6 and 48% when comparing 2008 and 2050 (see Fig. 6.8). On average, the efficiency of large appliances improves in a range from 21 to 48%, whereas the strongest decrease in specific energy demand can be seen for freezers. For ICT appliances, efficiency improves between 6 and 29% when comparing 2008 and 2050. The smallest efficiency improvement can be seen for televisions, due to compensating factors that increase energy demand such as the trend towards larger screens or new features (also known as the direct rebound effect). The ban on incandescent bulbs due to ecodesign requirements leads to their market phaseout. In parallel, halogen lamps increasingly become niche applications. The resulting market shares are filled by fluorescent lamps and light-emitting diodes (LED). In total, this improves the efficiency of lighting by 71% until 2050.

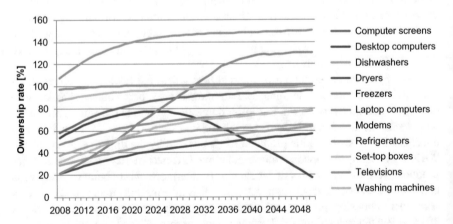

Fig. 6.7 Ownership rates of the major appliances for an average household in the EU 27 for the period 2008–2050

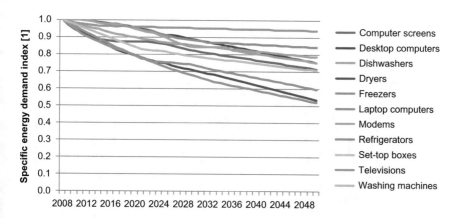

Fig. 6.8 Development of the specific energy demand of major appliances for an average house-hold in the EU 27 for the period 2008–2050

Comparing the internal heat gains related to end-use in 2008 indicates a broad range of 818–3929 kWh/dwelling (Fig. 6.9), which is equivalent to 3.8–6.6 W/m². In 2008, by far the highest internal heat gains related to end-use are found in Sweden (3929 kWh/dwelling) and Finland (3561 kWh/dwelling), whereas most Eastern European countries are on a level below 1500 kWh/dwelling. Analysing the results by end-use in 2008 reveals that the largest share is attributed to large appliances, lighting and the category 'Others'. In terms of growth until 2050, the largest variation is seen in Germany, Austria and Belgium countries with +22–36%. This is mainly attributed to a strong increase in ICT ownership rates and an ongoing rising trend of the category 'Others' as witnessed over the last two decades. In contrast, 14 countries witness a decrease of internal heat gains in the long-run either due to strong efficiency improvements or due to a large number of end-use already close to their saturation level in 2008 whereas moderate efficiency improvements are sufficient to overcompensate demand increasing effects.

In a first step the share of internal heat gains is analysed from an aggregated point of view. The results of the cohort modelling by reference building are cumulated on a country level to obtain an average SFH and MFH for 2008 and 2050 (Fig. 6.10). In addition to the country-specific results an average (AVG) per climate zone is given in Fig. 6.10, as a benchmark. The climate zones are distinguished by heating degree days (HDD) into a cold climate zone (>4200 HDD), a moderate climate zone (2200–4200 HDD) and a warm climate zone (<2200 HDD). On average, the specific useful energy demand of buildings declines by 47% until 2050 in case of a static approach, with cold climate zones witnessing the largest percentage decrease by about 54%, starting from a very high level in 2008 of 159 kWh/m²*a. Due to the improved thermal performance of buildings, the relative share of internal heat gains compared to the overall heat losses (transmission heat losses plus ventilation heat losses) also increases in the static norm-based approach

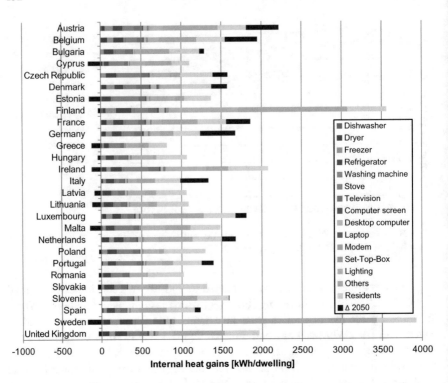

Fig. 6.9 End-use related annual internal heat gains per capita by country in 2008 and their change until 2050 (*dark grey area*)

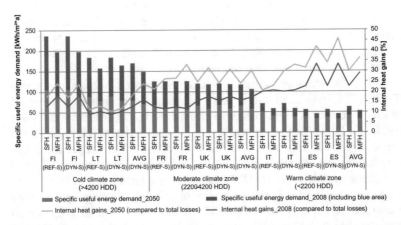

Fig. 6.10 Average useful energy demand and share of internal heat gains in selected EU27 countries in 2008 and 2050

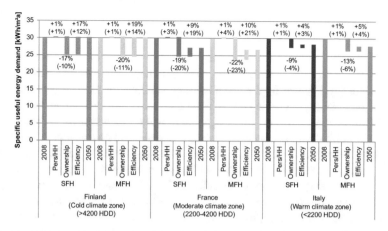

Fig. 6.11 Decomposition of useful energy demand for a nZEB (30 kWh/m²*a) of selected EU27 countries in 2008 and 2050

(REF-S), partially by a level of more than 50%. In terms of the DYN-S, the results indicate that the increasing share of internal heat gains is relatively larger for MFH than for SFH compared to the static approach. Especially in the warm climate zone the average share of internal heat gains (AVG) reaches a level of above 35% in 2050.

In a second step, a decomposition analysis is conducted for the dynamic approach to obtain deeper insights with regard to nZEB, for instance with a thermal performance of 30 kWh/m²*a. Figure 6.11 indicates that the increasing ownership rate of end-uses leads to increased internal heat gains which are cushioned by the decreasing number of persons per dwelling as well as the energy efficiency improvements of end-uses. The degree of compensation depends strongly on demographic change, the trend towards more single-person households and the ambitiousness of energy policies as well as their enforcement on a country basis. Comparing the countries in Fig. 6.11 with each other shows that climate-related heterogeneity as well as differences between SFH and MFH in terms of building physics lead to varying levels of internal heat gains. In Fig. 6.11, the values without brackets quantify the percentage impact of each effect on the overall change between 2008 and 2050 by selected country. The values in the brackets beneath represent the mean of each climate zone as a benchmark. Analysing the results shows that in Finland the demand decreasing effect of end-use ownership is almost compensated entirely by efficiency improvements. In contrast, the specific useful energy demand for France and Italy decreases by about 2 kWh/m²*a respectively 3 kWh/m²*a by 2050.

6.4 Conclusions

The analysis reveals that the values obtained from applying a static, norm-based approach systematically underestimate the contribution of internal heat gains to thermal heat demand. Comparing the benchmark evaluations for each climate zone indicates that, on average, the static and the dynamic share differ in a range from 20 to 70% by 2050. Increased ownership of appliances is the key driver for increased internal heat gains. Cushioning effects come from the decreasing number of persons per dwelling and increasing end-use efficiency. As the share of the heat load covered by internal heat gains correlates with a decrease in the specific useful energy demand, the magnitude of underestimation is amplified for a nZEB. In buildings with a specific useful energy demand of 30 kWh/m^2*a, for instance in France and Italy, the dynamic approach leads to a decrease in the specific useful energy demand by 10–15%. Overall, it can be concluded that it is essential to take dynamic heat dissipation into account for a differentiated evaluation of useful energy demand.

However, the results need to be interpreted with caution as they are essentially influenced by the assumptions of the underlying energy policy framework and the preferences of residents. In terms of the energy policy framework, the analysis revealed that the contribution of dynamic internal heat gains to cover useful energy demand is very sensitive to alternative developments of future minimum energy performance standards. Analysing alternative preference trajectories shows that the development of household ownership strongly depends on the assumed long-term saturation level, which is always based on expert guesses.

The analysis also contains elements of uncertainty due to limited empirical data. This is especially true for the utilization factor of end-uses and the average period residents spend in the heated building area. In particular, the contribution of the end-use category 'Others' is difficult to predict due to an increasing number of mobile devices. As 30–40% of the share of end-use-related internal heat gains can already be attributed to the category 'Others' in some countries, there is substantial uncertainty surrounding its future development when estimating internal heat gains.

The new modelling concept also provides elements for further research. At present, this concept is applied to analyse the interdependencies between technological, economic and building-physical parameters. However, in order to gain further insights into the contribution of internal heat gains to covering useful energy demand in residential buildings, a zoning by different types of heating areas within buildings could be applied combined with a more granular temporal resolution on an hourly basis.

Appendix

See Tables 6.2, 6.3 and 6.4.

Table 6.2 Minimum performance standards for refurbishment in the EU 27 by country [39]

Country	Roof	Wall	Floor	Window	Valid from
	Maximum U-value [W/m² K]				
Austria	0.20	0.35	0.40	1.40	2008
Belgium	0.30	0.40	0.40	2.50	2010
Cyprus	0.50	0.70	0.50	3.26	n. a.
Czech Republic	0.24	0.38	0.45	1.70	2009
Denmark	0.15	0.20	0.20	1.50	2006
Estonia	0.28	0.36	0.28	2.50	2003
Finland	0.14	0.14	0.09	1.00	2010
France	0.40	0.43	0.50	2.30	2007
Germany	0.24	0.24	0.30	1.30	2009
Greece	0.50	0.70	0.50	3.26	1980
Hungary	0.46	0.46	0.35	2.00	n. a.
Ireland	0.35	0.60	0.60	2.00	2006
Italy	0.34	0.42	0.43	2.73	2010
Latvia	0.20	0.25	0.25	1.80	2001
Lithuania	0.16	0.20	0.25	1.60	2005
Luxembourg	0.25	0.32	0.32	1.50	2008
Malta	0.34	0.42	0.43	2.73	n. a.
Netherlands	0.30	0.40	0.40	2.50	n. a.
Poland	0.30	0.30	n. a.	1.80	2008
Portugal	0.45	0.60	0.45	3.30	2006
Slovakia	0.46	0.46	0.35	2.00	2007
Slovenia	0.20	0.28	0.28	1.30	2008
Spain	0.54	0.97	0.66	4.20	2009
Sweden	0.13	0.18	0.15	1.30	2008
United Kingdom	0.16	0.30	0.25	2.00	2008
Romania	0.50	0.71	0.50	2.50	2007
Bulgaria	0.30	0.40	0.40	2.00	2005

Table 6.3 Subsidy programmes by country in the EU27 by country [16]

Country	Grants, subsidies	Loans	Tax incentives/levies
Austria	Existing and new buildings	No	Yes
Belgium	Existing and new buildings	No	Yes
Cyprus	Existing and new buildings	No	No
Czech Republic	Existing and new buildings	No	No
Denmark	Existing buildings	No	No
Estonia	No information available		
Finland	Existing and new buildings	No	Yes

(continued)

Table 6.3 (continued)

Country	Grants, subsidies	Loans	Tax incentives/levies
France	Existing and new buildings	Yes	Yes
Germany	Existing and new buildings	Yes	No
Greece	Existing buildings	No	Yes
Hungary	Existing buildings	No	Being planned
Ireland	Existing and new buildings	No	No
Italy	Existing buildings	Existing buildings	Yes
Latvia	Existing buildings	No	No
Lithuania	Existing and new buildings	No	No
Luxembourg	Existing and new buildings	New buildings	No
Malta	Existing and new buildings	No	No
Netherlands	Existing and new buildings	No	Yes
Poland	No subsidies	Existing buildings	No
Portugal	Existing and new buildings	No	Yes
Slovakia	Existing buildings	Existing buildings	No
Slovenia	Existing and new buildings	Yes	No
Spain	Existing and new buildings	Yes	Yes

Table 6.4 Overview of ecodesign product groups relevant for the residential sector (heating-related lots excluded) [40]

# Lot	Product group	Status quo
3	Desktops, laptops and computer screens	Regulations in force 2013. Draft ecodesign and labelling regulations for displays were discussed in the Consultation Forum 10 December 2014
5	Televisions	Draft ecodesign and labelling regulations for displays were be discussed in the Consultation Forum 10 December 2014
6	Standby and off-mode losses	Ecodesign regulation in force in January 2009
10	Room air-conditioning	Ecodesign regulation in force in March 2012. Labelling regulation in force in July 2011
10	Ventilation and kitchen hoods	Notification to the WTO in October 2013
13	Refrigerators and freezers	Ecodesign regulation in force 2009. Labelling regulation in force 2010
14	Washing machines	Regulation in force in 2010. Labelling regulation in force in 2010
14	Dishwasher	Ecodesign regulation in force 2010. Labelling regulation in force 2010
16	Dryer	Ecodesign and energy labelling regulations in force 2012
17	Vacuum cleaner	Ecodesign and labelling requirements in force in July 2013. Guidelines were published in 20 August 2014
18	Complex set-top-boxes	Voluntary agreement recognised
18	Simple set-top-boxes	Ecodesign requirements in force 25 February 2009
18	Directional lighting	Draft ecodesign regulation to WTO in 5 November 2014

References

1. Energy System Analysis Agency: Shaping Our Energy System Combining European Modelling Expertise. Karlsruhe (2013)
2. International Energy Agency: Energy Technology Perspectives: Scenarios and Strategies to 2050. Paris (2010)
3. Enerdata: Energy Efficiency Indicators in Europe. http://www.odyssee-indicators.org/. Grenoble. 30-1-2014
4. Deutsches Institut für Normung: DIN V 4108-6 - Wärmeschutz und Energie-Einsparung in Gebäuden, Teil 6: Berechnung des Jahresheizwärme- und des Jahresheizenergiebedarfs. Beuth Verlag, Berlin (2003)
5. Deutsches Institut für Normung: DIN V 4701-10 - Energetische Bewertung heiz- und raumlufttechnischer Anlagen, Teil 10: Heizung, Trinkwassererwärmung, Lüftung. Beuth Verlag, Berlin (2003)
6. Deutsches Institut für Normung: DIN EN ISO 13790—Calculation of Energy Use for Space Heating and Cooling. Beuth Verlag, Berlin (2008)
7. Bettgenhäuser, K.: Assessment modelling for building stocks—a technical, economical and ecological analysis. Doctoral thesis, Ingenieurwissenschaftlicher Verlag (2013)
8. IWU: Typology Approach for Building Stock Energy Assessment. Darmstadt (2012)
9. Kranzl, L., Hummel, M., Müller, M.A., Steinbach, J.: Renewable heating: perspectives and the impact of policy instruments. Energy Policy 59, 44–58 (2013)
10. Lee, T., Yao, R.: Incorporating technology buying behaviour into UK-based long term domestic stock energy models to provide improved policy analysis. Energy Policy 52, 363–372 (2013)
11. EUR-Lex: Directive 2009/125/EU—Establishing a Framework for the Setting of Eco-Design Requirements for Energy-Related Products (recast). Brussels (2009)
12. Elsland R, Boßmann T, Wietschel M.: Analysis of the European residential building stock until 2050—contribution of an energy-efficient building envelope to reduce space heating demand, KIC InnoEnergy scientist conference 2012
13. Elsland R, Schlomann B, Eichhammer W.: Is enough electricity being saved? Impact of energy efficiency policies addressing electrical household appliances in Germany until 2030. ECEEE Summer Study Proceedings 2013, pp. 1651–1662
14. Intelligent Energy Europe: ENTRANZE. http://www.entranze.enerdata.eu/. 17-11-2013
15. Intelligent Energy Europe (IEE): TABULA data and calculation workbook. http://www.building-typology.eu/downloads/public/calc/tabula-calculator.xls. 2-11-2012
16. Buildings Performance Institute Europe (BPIE): Data hub for the energy performance of buildings. http://www.buildingsdata.eu/. 21-10-2013
17. EUR-Lex: Directive 2010/31/EU—Energy Performance of Buildings (recast). Brussels (2010)
18. Weibull, W.: A statistical distribution function of wide applicability. J. Appl. Mech. 18, 293–297 (1951)
19. Tutz, G.: Regression for Categorial Data. Cambridge University Press, New York (2012)
20. Bass, F.: A new product growth for model consumer durables. Manage. Sci. 15, 215–227 (1969)
21. IWU: Der Einfluss des Gebäudestandards und des Nutzerverhaltens auf die Heizkosten. Darmstadt (2003)
22. Feist, W., Baffia, E., Schnieders, J., Pfluger, R.: Passivhaus Vorprojektierung. Darmstadt (2002)
23. Kildsgaard, I., Jarnehammar, A., Widheden, A., Wall, M.: Energy and environmental performance of multi-story apartment buildings built in timber construction using passive house principles. Buildings 3, 258–277 (2013)
24. Mata, M., Kalagasidis, A.S.: Calculation of Energy Use in the Swedish Housing. Goteborg (2009)

25. Monstvilas, E., Banionis, K., Stankevicius, V., Karbauskaite, J., Raimondas, B.: Heat gains in buildings—limit conditions for calculating energy consumption. J. Civ. Eng. Manage. **3**, 439–450 (2010)
26. Elsland, R., Peksen, I., Wietschel, M.: Are internal heat gains underestimated in thermal performance evaluation of buildings? Energy Procedia **62**, 32–41 (2014)
27. Ampatzi, E., Knight, I.: Modelling the effect of realistic domestic energy demand profiles and internal gains on the predicted performance of solar thermal systems. Energy Build. **55**, 285–298 (2012)
28. Jenkins, D., Liu, Y., Peacock, A.D.: Climatic and internal factors affecting future UK office heating and cooling energy consumptions. Energy Build. **40**, 874–881 (2008)
29. Firlag, S., Murray, S.: Impacts of airflows, internal heat and moisture gains on accuracy of modeling energy consumption and indoor parameters in passive building. Energy Build. **64**, 372–383 (2013)
30. Blight, T.S., Coley, D.: Sensitivity analysis of the effect of occupant behaviour on the energy consumption of passive house dwellings. Energy Build. **66**, 183–192 (2013)
31. Lubina, P., Nantka, M.B.: Internal Heat Gains in Relation to the Dynamics of Buildings Heat Requirements. Gliwice (2009)
32. Recknagel, H., Sprenger, E., Schramek, E.R.: Taschenbuch für Heizung und Klimatechnik. Oldenburg Verlag, Munich (2009)
33. Schweizerischer Ingenieur- und Architektenverein: SIA Norm 380/1: Thermische Energie im Hochbau. Zurich (2009)
34. Verein deutscher Ingenieure: VDI-Richtlinie 3804: Raumlufttechnik Bürogebäude. Dusseldorf (2009)
35. Sikula, O., Plasek, J., Hirs, J.: Numerical simulation of the effect of heat gains in the heating season. Energy Procedia **14**, 906–912 (2012)
36. Energy System Analysis Agency: ESA^2 web. http://www.esa2.eu/web/guest/mesap-web-access. Karlsruhe. 25-1-2014
37. EUR-Lex: Energy Performance of Buildings (recast). Directive 2009/31/EU. Brussels (2010)
38. Bundesministerium für Verkehr, Bau und Stadtentwicklung. Energetische Anforderungen und flankierende Maßnahmen für den Gebäudebestand in den mitteleuropäischen Nachbarländern. Hamburg (2010)
39. Hansen, P.: Auf dem Weg zum klimaneutralen Gebäudebestand in Europa bis 2050: Entwicklung der Energienachfrage, 7. Internationale Energiewirtschaftstagung an der TU Wien (IEWT 2011) (2011)
40. ECEEE: Products covered and their status in the ErP process. http://www.eceee.org/ecodesign/products. 02-02-2015

Chapter 7
Smart Home Appliance Control via Hand Gesture Recognition Using a Depth Camera

Dong-Luong Dinh and Tae-Seong Kim

Abstract The user-friendly and -intuitive interface for household appliances is considered as one of the highly promising fields for researches in the area of smart home and environment. Instead of traditional interface methodologies such as keyboard, mouse, touchscreen, or remote control, users in smart home or environment can control smart appliances via their hand gestures. This chapter presents a novel hand gesture interface system via a single depth imaging sensor to control smart appliances in smart home and environment. To control the appliances with hand gestures, our system recognizes the hand parts from depth hand silhouettes and generates control commands. In our methodology, we first create a database of synthetic hand depth silhouettes and their corresponding hand parts-labelled maps, and then train a random forests (RFs) classifier with the database. Via the trained RFs, our system recognizes the hand parts from depth silhouettes. Finally based on the information of the recognized hand parts, control commands are generated according to our predefined logics. With our interface system, users can control smart appliances which could be TV, radio, air conditioner, or robots with their hand gestures.

Keywords Hand gesture recognition · Depth imaging sensor · Smart appliances · Smart home · Smart environment

D.-L. Dinh
Department of Information Technology,
Nha Trang University, Nha Trang, Vietnam
e-mail: luongdd@ntu.edu.vn

T.-S. Kim (✉)
Department of Biomedical Engineering,
Kyung Hee University, Yongin, South Korea
e-mail: tskim@khu.ac.kr

© Springer International Publishing AG 2017
J. Littlewood et al. (eds.), *Smart Energy Control Systems for Sustainable Buildings*,
Smart Innovation, Systems and Technologies 67,
DOI 10.1007/978-3-319-52076-6_7

7.1 Introduction

Home is an important place of living for people especially the elderly and disabled. Home environment not only impacts on the quality of life, but it is also a place where people spend a large amount of their time. Recently, applying advanced technologies in various fields of architectural, electrical, automatic control, computer, and biomedical engineering to home is getting a lot of attentions from researchers to create smart home. One of the important technologies for smart home is how to control home environments. For instance, interaction via hand gesture is a more intuitive, natural, and intelligent way to interact with household appliances than the traditional interface methodologies using keyboard, mouse, touchscreen, or remote control devices, since users can interface with household appliances with just their hand gestures. Potential applications of such a human interaction based on hand gesture recognition include home entertainments [1], home appliances control [2], home healthcare systems [3, 4], etc. Among these applications, smart control for smart home appliances is one of the important applications for its daily usage: household appliances such as TV, radio, fans, and doors can be controlled by just hand gestures such as changing channels, temperature, and volume [5, 6].

Techniques for hand gesture recognition can be technically divided into two approaches: one is sensor-based and the other vision-based approaches [7] for static hand gestures which are typically represented by the orientation and shape of the hand pose in space at an instant of time without any movement and for dynamic hand gestures which include movement.

In the first approach, inertial sensors such as accelerometers or gyroscopes are typically used which are attached to hand parts to track and estimate their positions, velocity, and acceleration. Then motion features are extracted from measured signals and used for hand gesture recognition. In [8], the features were extracted from the angular velocity and acceleration of triaxial sensors and analysed by fisher discriminant analysis for 3D character gesture recognition. In [9, 10], the changes of accelerations in three perpendicular directions due to different gesture motions were used in real-time as features and then template matching and Hidden Markov Model (HMM) were employed to achieve gesture recognition. These studies had shown some success in recognizing hand sign gestures. However, most studies are focused on recognizing dynamic hand gestures based on motion features acquired through sensor signals, while static hand gesture or pose recognition are still remaining challenges due to motion sensor characteristics. Some applications of the gesture recognition technology in smart home environment have been developed in [11–13] where they used the sensor-based approach for dynamic hand gesture recognition to control home appliances such as radio, TV, and lightings. In these applications, the requirement of the sensor devices as a remote controller makes this kind approach unnatural and inconvenient in spite of its high sensitivity.

In the second approach, vision information from cameras is typically used for hand gesture recognition such as colour, shape, or depth. For static hand gesture or pose recognition which is done by extracting some geometric features such as fingertips, finger directions, and hand contours, or some non-geometric features such as such skin colour, silhouette, and texture. For instance, in [14], a static hand gesture recognition system was presented for nine different hand poses. Orientation and contour features were extracted for the hand gesture recognition. In [15], a real-time hand parts tracking technique was presented by using a cloth glove with various colours imprinted on it. The colour code and position features on the glove were used to track the hand parts for hand pose recognition. Meanwhile, for dynamic hand gesture recognition which is recognized by analyzing a consecutive sequence of recognized static hand poses. For instance, in [16], the key points were extracted as features from hand silhouettes using the Scale Invariance Feature Transform (SIFT) method. Their static hand gesture recognition method used bag-of-features and multiclass Support Vector Machine (SVM), and then a grammar was built to generate gesture commands from dynamic hand gestures to control an application or video game via hand gestures. Since the quality of image captured from RGB cameras is sensitive to the user environment such as noise, lighting conditions, and cluttered backgrounds. These studies have shown limited success in hand gesture recognition. Recently, with an introduction of new depth cameras, some studies used depth images for hand gesture recognition. For instance, in [17–19], the geodesic distances of depth map were utilized and considered as extracted features for hand gesture classification. In [20–23], a novel technique was presented which is one of the most popular and widely used methodologies for hand pose recognition, by directly recognizing hand parts: that is all pixels of given hand depth silhouette were assigned labels which were utilized to recognize hand parts. Based on the recognized parts, hand gestures were detected and recognized. Some applications to control home appliances in smart home using these techniques have been introduced. For example, in [24], an assisting system for the elderly and handicapped was developed to open or close household appliances such as TV, lamps, and curtains by hand gestures where hand poses was captured from three cameras and then recognized by getting their position and direction features. In [25], the authors used some extracted features from depth images which reflect the hand contour information to recognize some static hand poses. Then, dynamic hand gestures were recognized by considering a consecutive sequence of these static poses. Seven dynamic hand gestures were recognized and used for household appliances control.

In this chapter, we present a novel hand gesture recognition and Human Computer Interaction (HCI) system which recognizes each hand part in a hand depth silhouette and generate commands to control smart appliances in smart home environments. The main advantage of our proposed approach is that the state of each finger is directly identified by recognizing the hand parts and then hand gestures are recognized based on the state of each finger. We have tested and validated our system on real data. Our experimental tests achieved 98.50% in recognition of hand gestures with five subjects. Finally we have implemented and tested our HCI system through which one can control home appliances: smart home appliances can be turned on and off; channels and volumes can be changed with just simple hand gestures.

7.2 Hand Gesture-Based Interface System

The setting of our hand gesture recognition and HCI system for appliances control
in smart home environments is shown in Fig. 7.1. The system consists of two main
parts: a depth camera which is used to get hand depth silhouettes and an appliances
control interface which is used to give instructions to appliances. To make a user
friendly interface, our hand gesture interface system allows users to interface with
the appliances by understanding dynamic hand gestures which are recognized from
the hand poses and their movements as described in Tables 7.1 and 7.2.

Fig. 7.1 The setting of our
proposed hand gesture-based
interface system

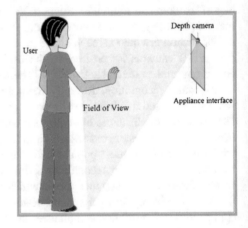

Table 7.1 Four basic types of hand poses

Pose ID	1	2	3	4
Function	Ready	Selection/enter	Exit	Move menu
Hand Poses				

Table 7.2 Types of static and dynamic hand gestures

Gesture types	Functionality of the system	Gesture types	Functionality of the system
1	Recognize Pose 1 to ready the interface system	4	Recognize and track Pose 1 as a hand mouse
2	Recognize Pose 2 to make a selection (i.e., enter)	5	Recognize and track Pose 4 to move the menu right or left
3	Recognize Pose 3 to exit the system	6	Recognize and track Pose 4 to turn up or down volume

Fig. 7.2 The framework of our hand gesture recognition system

7.3 Methodology

Overall process of our proposed system for hand gesture recognition, shown in Fig. 7.2, consists of two main parts: in the first part of hand parts recognition, a synthetic hand database (DB), which contains more than thousands of pairs of depth maps and their corresponding hand parts-labelled maps, was generated. Then, the DB was used in training Random Forests (RFs). In the recognition stage, a depth image was first captured from a depth camera and then a hand depth silhouette was extracted by removing the background. Next, the hand parts of a depth silhouette were recognized using the trained RFs. In the second part of hand gesture recognition, a set of features was extracted from the labelled hand parts. Finally, based on the extracted features, hand gestures were recognized by our rules, generating interface commands.

7.3.1 Hand Depth Silhouette Acquisition

In our work, we used a creative interactive gesture camera [26]. This device is capable of close-range depth data acquisition. The depth imaging parameters were set with the image size of 240×320, and frame rate of 30 fps. The hand parts were captured in the field of view of $70°$.

To detect the hand area and remove background, we used an adaptive depth threshold. The value of threshold was determined based on a specific distance from a depth camera to the hand parts. Hand depth silhouettes were extracted with the background removal methodology mentioned in [26]. The detected and segmented hand is shown in Fig. 7.3.

Fig. 7.3 A segmented hand
depth silhouette

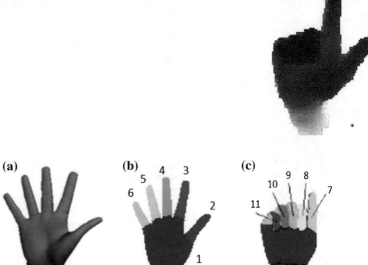

(a)   **(b)**   **(c)**

Fig. 7.4 Hand model: **a** 3D hand model in 3Ds-max and **b, c** hand parts with twelve labels

7.3.2 Hand Parts Recognition

a. Synthetic hand DB generation

To recognize hand parts from a hand depth silhouette via RFs, the synthetic hand
DB, which contains pairs of depth images and their corresponding hand part-labelled
maps, is needed to train RFs. We created the DB with a synthetic hand model using
3Ds Max, a commercial 3D graphic package [27]. To identify hand parts, twelve
labels were assigned to each hand model as shown in Fig. 7.4 and Table 7.3. The
five fingers including the thumb, index, middle, ring, and pinkie fingers, were represented by ten corresponding labels including the five front and five back sides of
the fingers. The front parts were coded with the indices of 2, 3, 4, 5 and 6. Likewise,
the five back sides were coded with the indices of 7, 8, 9, 10 and 11, respectively.
The images in the DB had a size of 320×240 with 16-bit depth values.

b. Depth feature extraction

To extract depth features f from pixel p of depth silhouette I as described in [28], we
computed a set of depth features of the pixel p based on difference between the
depth values of a neighbourhood pixel pair in depth silhouette I. The positions of
the pixel pairs were randomly selected on the depth silhouette and they had a
relation with the position of the considered pixel p by two terms o_1 and o_2 of the
coordinates of x and y, respectively. Depth features, f, are computed as follows:

Table 7.3 A list of the named and labelled parts in hand model

Label no.	Hand parts	Label no.	Hand parts
1	Palm	7	Thumb (back)
2	Thumb (front)	8	Index (back)
3	Index (front)	9	Middle (back)
4	Middle (front)	10	Ring (back)
5	Ring (front)	11	Pinkie (back)
6	Pinkie (front)	12	Wrist

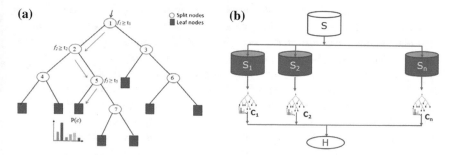

Fig. 7.5 RFs for pixel-based classification: **a** a binary decision tree and **b** ensemble of n decision trees for training RFs

$$f(I,p) = I_d\left(x_p + \frac{o_{1x}}{I_d(x_p, y_p)}, y_p + \frac{o_{1y}}{I_d(x_p, y_p)}\right)$$
$$- I_d\left(x_p + \frac{o_{2x}}{I_d(x_p, y_p)}, y_p + \frac{o_{2y}}{I_d(x_p, y_p)}\right), \qquad (7.1)$$

where $I_d(x_P, y_P)$ is the depth value at the coordinates of (x_P, y_P) and (o_1, o_2) is an offset pair. The maximum offset value of o_1, o_2 pairs was 30 pixels that corresponds to 0.6 meters which was the distance of the subject to the camera. The normalization of the offset by $\frac{1}{I_d(x_p,y_p)}$ ensured that the features are distance invariant.

c. Hand part recognition via RFs

In our work, we utilized RFs for hand parts recognition. RFs are a combination of tree predictors such that each tree depends on the values of a random vector sampled independently with the same sample distribution for all trees in the forest [29]. These concepts are illustrated in Fig. 7.5. Figure 7.5a presents a single decision tree learning process as a tree predictor. The use of a multitude of decision trees for training and testing RFs on the same DB S is described in Fig. 7.5b. The sample sets $\{S_i\}_{i=1}^{n}$ are drawn randomly from the training data S by bootstrap algorithm [29].

In training, we used an ensemble of 21 decision trees. The maximum depth of trees was 20. Each tree in RFs was trained with different pixels sampled randomly from the DB. A subset of 500 training sample pixels was drawn randomly from

Fig. 7.6 An example of labelled hand parts recognition: **a** a hand depth silhouette and **b** the labelled hand parts

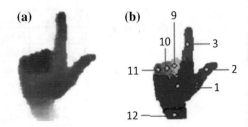

each synthetic depth silhouette. A sample pixel was extracted as in Eq. (7.1), to obtain 800 candidate features. At each splitting node in the tree, a subset of 28 candidate features was considered. For pixel classification, each pixel p of a tested depth silhouette was extracted to obtain the candidate features. For each tree, starting from the root node, if the value of the splitting function was less than a threshold of the node, p went to left and otherwise p went to right. The optimal threshold for splitting the node was determined by maximizing the information gain in the training process. The probability distribution over 12 hand parts was computed at the leaf nodes in each tree. Final decision to label each depth pixel for a specific hand part was based on the voting result of all trees in the RFs.

To recognize hand parts of each hand depth silhouette, all pixels of each hand depth silhouette were classified by the trained RFs to assign a corresponding label out of the 12 indices. A centroid point was withdrawn from each recognized hand part, representing each hand part as illustrated in Fig. 7.6.

7.3.3 Hand Gesture Recognition

a. Hand poses recognition

From the recognized hand parts, we extracted a set of features. In our labelling, each finger was represented by two different labels: one label for its front side corresponding to the open state of the finger and another for its back side corresponding to the close state of the finger. From the information of the recognized hand parts, we identify the open or close states of each finger. The states of the five labelled fingers were identified and saved as features, namely f_{Thumb}, f_{Index}, f_{Middle}, f_{Ring}, and f_{Pinkie} respectively.

$$f_{Fingers}(i) = \begin{cases} 1 : Open & for \quad Label = 2, 3, 4, 5, \, or \, 6 \\ 0 : Close & for \quad Label = 7, 8, 9, 10, \, or \, 11 \end{cases} \tag{7.2}$$

For example, as shown in Fig. 7.6b, f_{Thumb} and f_{Index} become 1 corresponding to the open state of the fingers. In contrast, f_{Middle}, f_{Ring}, and f_{Pinkie} become 0 corresponding to the close state of the fingers.

Table 7.4 Recognition rules of the four basic hand poses based on the states of the five fingers

f_{Thumb}	f_{Index}	f_{Middle}	f_{Ring}	f_{Pinky}	Pose type	Hand poses
0	1	0	0	0	1	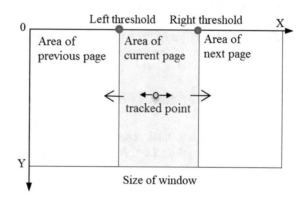
1	1	0	0	0	2	
0	0	0	0	0	3	
1	1	1	1	1	4	

0 = Close and 1 = Open

Fig. 7.7 Three divided areas in the screen window to recognize dynamic hand gestures (i.e., Gesture types 5, 6, and 7)

To recognize four basic hand poses, we derived a set of recognition rules. The set of five features from the states of all fingers was used to decode the meaning of the four hand poses. The derived recognition rules are given in Table 7.4.

b. Hand gesture recognition

To understand hand gestures with the recognized hand poses as explained in Table 7.2, after recognizing the hand poses, we tracked their positions. To understand Gestures 1, 2, and 3, our system recognizes Pose 1, 2, and 3, respectively. Then Gesture 4 can be understood by recognizing Pose 1 and then tracking the centroid point position of the index finger in the x and y dimension which gets mapped on the interface screen: it acts as a hand mouse. To understand Gesture 5 which is used to move the menu to the right or left, Pose 4 is recognized and the centroid point position of the palm is tracked in the x dimension between two consecutive frames including the previous and current frames. By dividing the frame window into three sub-areas as presented in Fig. 7.7, if the tracked point of the current frame is moved from the area of current or previous pages to the area of next page, the screen menu slides to the right. Likewise, if the tracked point of the current frame is moved from the area of current or next pages to the area of previous page the screen menu slides to the left. To understand Gesture 6 to turn up or down volume, Pose 4 is recognized and then the centroid point position of the palm is tracked in the y dimension. The difference between the tracked points position of two consecutive frames is used to turn up or down the volume.

7.4 Experimental Results and Demonstrations

7.4.1 Results of Hand Parts Recognition

To evaluate our hand parts recognition quantitatively, we tested on a set of 500 hand depth silhouettes containing various poses over the four hand poses. The average recognition rate of the hand parts was 96.90%. Then, we assessed the hand parts recognition on real data qualitatively, since the ground truth labels are not available. We only performed visual inspections on the recognized hand parts. A representative set of the recognized hand parts are shown in Fig. 7.8.

7.4.2 Results of Hand Pose Recognition

To test our proposed hand poses recognition methodology, a set of hand depth silhouettes was acquired from five different subjects. Each subject was asked to

Fig. 7.8 A set of representative results of the recognized hand parts via the trained RFs on real data

make 40 hand poses. Table 7.5 shows the recognition results of the four hand poses in a form of confusion matrix. The mean recognition rate of 98.50% was achieved.

7.4.3   Graphic User Interfaces (GUIs) for Demonstrations

We designed and implemented a GUI interface of our hand gesture interface system as presented in Figs. 7.9 and 7.10. How to use of our system can be explained by the following two examples. In example 1, to open and select any channel for TV, one should use Gesture 1 to open (i.e., activate) the system, use Gesture 4 as a hand mouse to select the TV icon on the GUI screen in Fig. 7.6a, and use Gesture 2 to open the TV. Then the TV channels as shown in Fig. 7.6b opens for a selection of a

Table 7.5 The confusion matrix of the four recognized hand poses using our proposed system	Pose types	1	2	3	4
	1	**100**			
	2		**100**		
	3		2	**98**	
	4	1	1	2	**96**

(a)   (b)

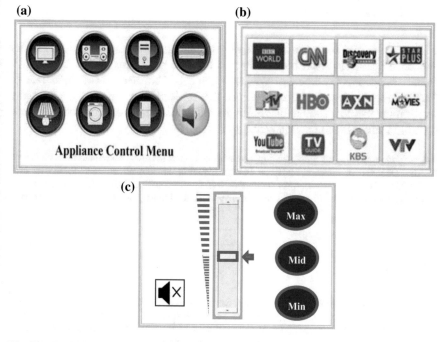

(c)

Fig. 7.9 Illustrated gesture-based GUIs of appliance control system. The GUIs for **a** appliances selection, **b** TV channel selection, and **c** volume control

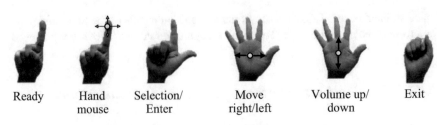

| Ready | Hand mouse | Selection/ Enter | Move right/left | Volume up/ down | Exit |

Fig. 7.10 A hand gestures sequence and their tracked points used in our system

channel. To browse the channel menus, one use Gesture 5 to slide the channel pages to the right (i.e., the next page) or left (i.e., the previous page), use Gestures 4 and 2 to select a channel, and finally use Gesture 3 to get back to the main GUI menu as shown in Fig. 7.6a. In example 2, to control volume as well as change temperature, intensity of lightings, or speed of fans, one should use Gesture 1 to open the system, use Gesture 4 as a hand mouse to select the volume icon on the GUI screen, and then use Gesture 2 to open the volume screen as shown in Fig. 7.6c. To change volume, use Gesture 6 to turn up or down volume or use Gestures 4 and 2 to select the fixed minimum (Min), middle (Mid), or maximum (Max) volume levels. To get back to the main GUIs, one can use Gesture 3.

7.5 Conclusions

In this work, we have presented a novel hand gesture recognition system for appliance control in smart home using the labelled hand parts via the trained RFs from a hand depth silhouette. We have achieved the mean recognition rate of 98.50% over the four hand gestures from five subjects. Our proposed hand gesture recognition method should be useful in automation applications for appliance control in smart home environment.

Acknowledgements This work was supported by the Center for Integrated Smart Sensors funded by the Ministry of Science, ICT & Future Planning as Global Frontier Project (CISS-2011-0031863). This work was supported by International Collaborative Research and Development Programme (funded by the Ministry of Trade, Industry and Energy (MOTIE, Korea) (N0002252).

References

1. Reifinger, S., Wallhoff, F., Ablassmeier, M., Poitschke, T., Rigoll, G.: Static and dynamic hand-gesture recognition for augmented reality applications. In: Human-Computer Interaction. HCI Intelligent Multimodal Interaction Environments, pp. 728–737. Springer, Berlin (2007)
2. Shao, J., Ortmeyer, C., Finch, D.: Smart home appliance control. In: Industry Applications Society Annual Meeting, 2008. IAS'08, IEEE, pp. 1–6 (2008)

3. Bien, Z.Z., Park, K.H., Jung, J.W., Do, J.H.: Intention reading is essential in human-friendly interfaces for the elderly and the handicapped. IEEE Trans. Ind. Electron. **52**(6), 1500–1505 (2005)
4. Levin-Sagi, M., Pasher, E., Carlsson, V., Klug, T., Ziegert, T., Zinnen, A.: A comprehensive human factors analysis of wearable computers supporting a hospital ward round. In: 2007 4th International Forum on Applied Wearable Computing (IFAWC), pp. 1–12 (2007)
5. Shimada, A., Yamashita, T., Taniguchi, R.I.: Hand gesture based TV control system— towards both user- and machine-friendly gesture applications. In: 2013 19th Korea-Japan Joint Workshop on Frontiers of Computer Vision, (FCV), IEEE, pp. 121–126 (2013)
6. Bhuiyan, M., Picking, R.: Gesture-controlled user interfaces, what have we done and what's next. In: Proceedings of the Fifth Collaborative Research Symposium on Security, E-Learning, Internet and Networking (SEIN 2009), pp. 25–29. Darmstadt, Germany (2009)
7. Murthy, G.R.S., Jadon, R.S.: A review of vision based hand gestures recognition. Int. J. Inf. Technol. Knowl. Manage. **2**(2), 405–410 (2009)
8. Oh, J.K., Cho, S.J., Bang, W.C., Chang, W., Choi, E., Yang, J., Kim, D.Y.: Inertial sensor based recognition of 3D character gestures with an ensemble classifiers. In: Ninth International Workshop on Frontiers in Handwriting Recognition, 2004. IWFHR-9 2004, IEEE, pp. 112–117 (2004)
9. Zhou, S., Shan, Q., Fei, F., Li, W.J., Kwong, C.P., Wu, P.C., Liou, J.Y.: Gesture recognition for interactive controllers using MEMS motion sensors. In: 4th IEEE International Conference on Nano/Micro Engineered and Molecular Systems, 2009. NEMS 2009, IEEE, pp. 935–940 (2009)
10. Xu, R., Zhou, S., Li, W.J.: MEMS accelerometer based nonspecific-user hand gesture recognition. Sens. J. IEEE **12**(5), 1166–1173 (2012)
11. Mäntyjärvi, J., Kela, J., Korpipää, P., Kallio, S.: Enabling fast and effortless customisation in accelerometer based gesture interaction. In: Proceedings of the 3rd International Conference on Mobile and Ubiquitous Multimedia, pp. 25–31. ACM (2004)
12. Wan, S., Nguyen, H.T.: Human computer interaction using hand gesture. In: 30th Annual International Conference of the IEEE, Engineering in Medicine and Biology Society, 2008. EMBS 2008, IEEE, pp. 2357–2360 (2008)
13. Ng, W.L., Ng, C.K., Noordin, N.K., Ali, B.M.: Gesture based automating household appliances. In: Human-Computer Interaction. Interaction Techniques and Environments, pp. 285–293. Springer, Berlin (2011)
14. Vieriu, R.L., Goras, B., Goras, L.: On HMM static hand gesture recognition. In: 2011 10th International Symposium on Signals, Circuits and Systems (ISSCS), IEEE, pp. 1–4 (2011)
15. Wang, R.Y., Popović, J.: Real-time hand-tracking with a color glove. In: ACM Transactions on Graphics (TOG), vol. 28, no. 3, p. 63. ACM (2009)
16. Dardas, N.H., Georganas, N.D.: Real-time hand gesture detection and recognition using bag-of-features and support vector machine techniques. IEEE Trans. Instrum. Measur. **60**(11), 3592–3607 (2011)
17. Plagemann, C., Ganapathi, V., Koller, D., Thrun, S.: Real-time identification and localization of body parts from depth images. In: 2010 IEEE International Conference on Robotics and Automation (ICRA), IEEE, pp. 3108–3113 (2010)
18. Molina, J., Escudero-Viñolo, M., Signoriello, A., Pardàs, M., Ferrán, C., Bescós, J., Martínez, J.M.: Real-time user independent hand gesture recognition from time-of-flight camera video using static and dynamic models. Mach. Vis. Appl. **24**(1), 187–204 (2013)
19. Liang, H., Yuan, J., Thalmann, D., Zhang, Z.: Model-based hand pose estimation via spatial-temporal hand parsing and 3D fingertip localization. Visual Comput **29**(6–8), 837–848 (2013)
20. Kang, B., Rodrigue, M., Hollerer, T., Lim, H.: Poster: real time hand pose recognition with depth sensors for mixed reality interfaces. In: 2013 IEEE Symposium on 3D User Interfaces (3DUI), IEEE, pp. 171–172 (2013)

21. Keskin, C., Kıraç, F., Kara, Y. E., Akarun, L.: Real time hand pose estimation using depth sensors. In: Consumer Depth Cameras for Computer Vision, pp. 119–137. Springer, London (2013)
22. Luong, D.D., Lee, S., Kim, T.S.: Human computer interface using the recognized finger parts of hand depth silhouette via random forests. In: 2013 13th International Conference on Control, Automation and Systems (ICCAS), IEEE, pp. 905–909 (2013)
23. Zhao, X., Song, Z., Guo, J., Zhao, Y., Zheng, F.: Real-time hand gesture detection and recognition by random forest. In: Communications and Information Processing, pp. 747–755. Springer, Berlin (2012)
24. Bien, Z.Z., Do, J.H., Kim, J.B., Stefanov, D., Park, K.H.: User-friendly interaction/interface control of intelligent home for movement-disabled people. In: Proceedings of the 10th International Conference on Human-Computer Interaction (2003)
25. Wu, C.H., Lin, C.H.: Depth-based hand gesture recognition for home appliance control. In: 2013 IEEE 17th International Symposium on Consumer Electronics (ISCE), IEEE, pp. 279–280 (2013)
26. www.intel.com/software/perceptua
27. Autodesk 3Ds MAX, 2012
28. Shotton, J., Sharp, T., Kipman, A., Fitzgibbon, A., Finocchio, M., Blake, A., Moore, R.: Real-time human pose recognition in parts from single depth images. Commun. ACM 56(1), 116–124 (2012)
29. Hastie, T., Tibshirani, R., Friedman, J., Hastie, T., Friedman, J., Tibshirani, R.: The Elements of Statistical Learning, vol. 2, no. 1, Springer, New York (2009)

Chapter 8
Neural Networks Applied to Short Term Load Forecasting: A Case Study

Filipe Rodrigues, Carlos Cardeira and J.M.F. Calado

Abstract A good management of renewable energy systems and energy storage requires Short Term Load Forecasting (STLF). In particular, Artificial Neural Networks (ANN) have proved their ability to cope with data driven nonlinear models. In this paper ANN models are used with input variables such as apartment area, numbers of occupants, electrical appliance consumption and time, in order to achieve a robust model to be used in forecasting energy consumption of general homes. A feed-forward ANN trained with the Levenberg-Marquardt algorithm is tested and their results show a quite accurate model foreseeing that ANNs are a promising tool for STLF.

Keywords Artificial neural networks · Levenberg-Marquardt · Energy forecasting · Hourly and daily energy · Boolean application

F. Rodrigues (✉) · J.M.F. Calado
Departamento de Engenharia Mecânica, IDMEC/ISEL—Instituto
Superior de Engenharia de Lisboa, Instituto Politécnico de Lisboa,
Rua Conselheiro Emídio Navarro, 1959-007 Lisbon, Portugal
e-mail: fmrodrigues@dem.isel.pt

J.M.F. Calado
e-mail: jcalado@dem.isel.ipl.pt

F. Rodrigues
MIT Portugal, Campus IST-Tagus Park, Av. Professor Cavaco Silva,
2744-016 Porto Salvo, Portugal

C. Cardeira · J.M.F. Calado
IDMEC/LAETA, Instituto Superior Técnico, Universidade de Lisboa,
Av. Rovisco Pais, 1049-001 Lisbon, Portugal
e-mail: carlos.cardeira@tecnico.ulisboa.pt

© Springer International Publishing AG 2017   173
J. Littlewood et al. (eds.), *Smart Energy Control Systems for Sustainable Buildings*,
Smart Innovation, Systems and Technologies 67,
DOI 10.1007/978-3-319-52076-6_8

8.1   Introduction

The increased of electricity demand, dioxide carbon emission and the energy resource scarcity are a worldwide concern. This paper seeks to contribute to the sustainable development concept, the rational use and storage of electric power.

The aim of this paper is to show the applicability of an Artificial Neural Network (ANN) approach to develop a simple and reliable method to forecast households' daily and hourly energy consumption for residential and small buildings. ANN favored as a method for forecasting in many applications.

Short-term forecasts, in particular, have become increasingly important since the rise of the competitive energy markets. Most forecasting models and methods have already been tried out on load forecasting, with varying degrees of success.

In contrast with the majority of the works published in this area, this work does not make use of temperature as a network input. Forecasting is made using only past load values and hour of the day. Our objective is to show that if an adequate training set is chosen, it is possible to obtain good results even without considering a huge amount of data like temperature or any other weather data.

The increasing use of renewable energy resources, namely the use of wind power in a larger scale, leads to issues of variable production rates that do not necessary fit to the demand of the consumers. Forecasting the consumers demand becomes of the utmost importance because some of the non-urgent consumer needs of energy (e.g. the turn on or off of the internal cycles of washing machines, air conditioning and fridge equipment) may be relatively shifted (delayed or anticipated) to achieve a better fit of the production profile to the consumption profile without compromising the comfort of the consumer and services level. The use of energy accumulators is always needed but their size can be reduced if a correct forecast of the energy consumption is available.

This work is dedicated to tackle the problem of energy consumption forecasts based in a 18 months long comprehensive set of data obtained from monitoring the real households energy consumption with a 15 mn granularity.

The ANN have been trained and tested using electricity energy consumption data obtained from direct logged in 93 households. A database with consumption records measured between February 2000 and July 2001, in Portugal that was used by Sidler's research team [1], has been used during the current studies. For Short Term Electric Load Forecasting (STELF), as the daily and hourly consumption forecast, this research used inputs that include apartment's areas, inhabitants and electricity energy consumption appliances data, and the daily or the hourly load profile electricity consumption of each household was the output. After the selecting and accept the valid data, all of the daily and hourly energy consumption values of 18 months were used in training and testing the model.

The research analyzed several models, identified and concluded that using MLP neural networks with backpropagation, Levenberg-Marquardt learning algorithm, hyperbolic tangent activation function in input and output layers and 20 neurons in the hidden layer, reveals to be a worthy method to provide acceptable results. For each network, fraction of variance (R^2), mean absolute percent error (MAPE) and the standard deviation of error (SDE) values were calculated and compared.

This paper is organized as follows: the second section presents the ANN architecture adopted; the third and fourth section explains the load forecasting models, their training processes and provides a deeper discussion about the results achieved. The last section presents the main conclusions and indicates some directions for future work.

8.2 State of the Art

The ability to forecast electricity load requirements is one of the most important aspects of effective management of power systems. The quality of the forecasts directly impacts the economic viability and the reliability of every electricity company. Many important operating decisions such as scheduling of power generation, scheduling of fuel purchasing, maintenance scheduling, and planning for energy transactions are based on electric load forecasting.

There are three different types of electric load forecasting depending on the time horizon and the operating decision that needs to be made: short term, medium term, and long term forecasting. In general, long term forecasting is needed for power system planning, medium term forecasting is needed for maintenance and fuel supply planning, while short term forecasting is needed for the day to day operation of the power system.

The technical literature is abundant with techniques and approaches for performing or improving short term load forecasting. A number of approaches work well with certain power systems or certain geographical areas, while they fail for some other systems due to the nature of the electric load demand: it is complex, highly nonlinear, and dependent on weather, seasonal and social factors.

A number of researchers have compiled extensive surveys on load forecasting. Some of these surveys have been focused on neural networks for short term load forecasting (STLF) [2, 3], some other techniques used such as time series and regression models [4], as well as approaches based on exponentially weighted methods [5], while some other authors provided a general look at all types of load forecasting methodologies [6]. Artificial Neural Networks (ANN) have received a large share of attention and interest. Neural networks are probably the most widely used Artificial Intelligence (AI) technique in load forecasting and papers in the area that have been published since the late 80s. The two main reasons for the use of neural networks as a tool for load forecasting is the fact that neural networks can approach numerically any desired function and the fact of not being dependent upon models. Most studies forecasting with neural networks can be divided into forecasts of an output variable to forecast short-term or multiple outputs to forecast the load curve.

The review of literature mentioned in this paper enabled us to establish in our research a number of variables that are based on the rise in consumption patterns by households, both during weekdays and weekend, and use the ANN to make forecast the typical values for energy consumption per household for one day, considering a specific type of household.

AlFuhaid et al. [7] use a cascaded neural network for predicting load demands for the next 24 h.

Aydinalp et al. [8] introduced a comprehensive national residential energy consumption model using an ANN methodology. They divided it into three separate models: appliances, lighting and cooling (ALC); Domestic hot water (DHW); and space heating (SH).

Becalli et al. [9] describe an application based on ANN to forecast the daily electric load profiles of a suburban area, using a model based on a multi-layer perceptron (MLP), having as inputs load and weather data. Few years later, Beccali et al. [10] used a forecasting model based on an Elman recurrent neural network to obtain lower prediction error rates assessing the influence of Air Conditioning systems on the electric energy consumption of Palermo, in Italy. The model estimates the electricity consumption for each hour of the day, starting from weather data and electricity demand related to the hour before the hour of the forecast.

Tso and Yau [11] present three modeling techniques for the prediction of electricity energy consumption in two year seasons: summer and winter. The three predictive modeling techniques are multiple regressions, a neural network and decision tree models. When comparing accuracy in predicting electricity energy consumption, it was found that the decision tree model and the neural network approach perform slightly better than the other modeling methodology in the summer and winter seasons, respectively.

Despite the conditional demand analysis (CDA) method [12] is capable of accurately predict the energy consumption in the residential sector as well as others [13–16] at the regional level and national level, however, considering households by households, it was shown that the CDA model has limited utility for modeling the energy consumption in the residential sector.

ANNs are reliable as a forecasting method in many applications, however load forecasting is a difficult task. First, because the load series are complex and exhibits several levels of seasonality: the load at a given hour is dependent not only on the load at the previous hour, but also on the load at the same hour on the previous day, and on the load at the same hour on the day with the same denomination in the previous week. Secondly, there are many important exogenous variables that must be considered, especially weather-related variables. Most authors run their simulations using logged weather values instead of forecasted ones, which is a standard practice in electric load forecasting. However, one should take into account that the forecasting errors in practice will be larger than those obtained in simulations, because of the added weather forecast uncertainty [2].

The work described in this paper was carried out with the aim to achieve forecast approaches for daily and hourly energy consumption of a random household and prediction of several days' energy consumption, using ANNs and a Boolean metering system. The results achieve are encouraging revealing that ANNs are able to forecast daily and hourly energy consumption, as well as a reliable load profile.

To model the electric load profile, it has been used as ANNs training data, hourly and daily data measured at end-use energy consumptions. The energy consumption and electric load profiles have been recorded considering several weeks, including both weekdays and weekend.

8.3 Methodology

8.3.1 Designing Artificial Neural Networks

Selecting an appropriate architecture is in general the first step to take when designing an ANN-based forecasting system. In the current studies, the network architecture was built based on multilayer perceptron (MLP), full-connected, which is a feed-forward type of artificial neural network, and the training task was performed through a backpropagation learning algorithm.

The backpropagation learning algorithms, most common in use in feed-forward ANN, are based on steepest-descent methods that perform stochastic gradient descent on the error surface. Backpropagation is typically applied to multiple layers of neurons and works by calculating the overall error rate of an artificial neural network. The output layer is then analyzed to see the contribution of each of the neurons to that error. The neurons weights and threshold values are then adjusted, according to how much each neuron contributed to the error, to minimize the error in the next iteration. To customize its output, several backpropagation learning algorithms use the Momentum parameter. Momentum can be useful to prevent a training algorithm from getting trapped in a local minimum and determines how much influence the previous learning iteration will has on the current iteration.

In this study, it was used the learning Levenberg-Marquardt algorithm that works as a training algorithm with the capabilities of the pruning methodologies. Pruning is a process of examining a solution network, determining which units are not necessary to the solution and removing those units [17]. By doing the artificial neural network prune the model achieved has reduced complexity and the computational effort to run it, especially in real time, is reduced too.

While backpropagation is a steepest descent algorithm, the Levenberg-Marquardt algorithm is a variation of Newton's method. The advantage of Gauss–Newton over the standard Newton's method is that it does not require calculation of second-order derivatives. The Levenberg-Marquardt algorithm trains an ANN faster (10–100 times) than the usual backpropagation algorithms. The Levenberg-Marquardt algorithm, which consists in finding the update given by:

$$\Delta x = -[J^T(x) J(x) + \mu I]^{-1} J^T(x) e(x) \tag{8.1}$$

where $J(x)$ is the Jacobian matrix, μ is a parameter conveniently modified during the algorithm iterations and $e(x)$ is the error vector. When μ is very small or null the Levenberg-Marquardt algorithm becomes Gauss-Newton, which should provide faster convergence, however for higher μ values, when the first term within square brackets of Eq. (8.1) is negligible with respect to the second term within square brackets, the algorithm becomes steepest descent. Hence, the Levenberg-Marquardt algorithm provides a nice compromise between the speed of Gauss-Newton and the guaranteed convergence of steepest descent [18].

Therefore, the Levenberg-Marquardt (LM) algorithm combines the robustness of the steepest-descent method with the quadratic convergence rate of the Gauss–Newton method to effectively identify the minimum of a convex objective function. LM outperforms gradient descent and conjugate gradient methods for medium sized nonlinear models. It is a well optimization procedure and it is one that works extremely well in practice. For medium sized networks (a few hundred weights say) this method will be much faster than gradient descent plus momentum and even allows to achieve a model with reduced complexity.

Since one hidden layer is enough to approximate any continuous functions [2], the MLP used in the current studies has three layers: an input layer, a hidden layer and an output layer. The number of neurons used in the input layer depends on the number of input variables being considered as inputs of the forecasting model, while the number of neurons in the output layer depends on the number of variables that one wants to predict with a specific model. Furthermore, the number of neurons in the hidden layer comes down by trial and error procedure to determine a workable number of hidden neurons to use. Selecting a few alternative numbers and then running simulations to find out the one that gave the best fitting (or predictive) performance allows to achieve the number of neurons in the hidden layer.

The activation function for the hidden and output neurons must be differentiable and non-decreasing [2]: as mentioned below in the current studies has been adopted the linear and the hyperbolic tangent functions. The activation functions are used to scale the output given by the neurons in the output layer.

As mentioned above, the ANN architecture was built with three-layer feedforward configuration. In order to achieve the desired performance for the forecasting model obtained, as well as during the artificial neural network training stage, following a trial and error procedure, the network was optimized by varying properties such as learning algorithm, scaling μ, and number of neurons, being the performance evaluated by maximizing the R^2 and minimizing the mean square error (MSE) values. Once defined the final architecture the selected learning algorithm was used during several training cycles to achieve the final connections weights and the bias values. The next step was concerned with test the performance of the model achieved.

McCullouch and Pitts [19] created a computational model for neural networks based on mathematics and algorithms. They called this model threshold logic. The model paved the way for neural network research to split into two distinct approaches. One approach focused on biological processes in the brain and the other focused on the application of neural networks to artificial intelligence.

In the late 1940s psychologist Hebb [20] created a hypothesis of learning based on the mechanism of neural plasticity that is now known as Hebbian learning. Hebbian learning is considered to be a 'typical' unsupervised learning rule and its

later variants were early models for long term potentiation. These ideas started
being applied to computational models in 1948 with Turing's B-type machines.

Farley and Clark [21] first used computational machines, then called calculators,
to simulate a Hebbian network at MIT. Other neural network computational
machines were created by Rochester et al. [22] and nowadays there are several
companies developing software to use ANN, as Mathworks leading developer of
mathematical computing software for engineers and scientists, founded in 1984,
who has Matlab ANN toolbox.

Figure 8.1 shows a neuron with its external inputs, connections and output.

The MATLAB neural network toolbox (nntool) has been used to train the
feedforward ANNs. MATLAB provides built-in transfer functions that have been
used for the hidden and output layers as follows: hyperbolic tangent sigmoid
(tansig) for the hidden neurons; a pure linear function (purelin) for the output
neurons (Fig. 8.2).

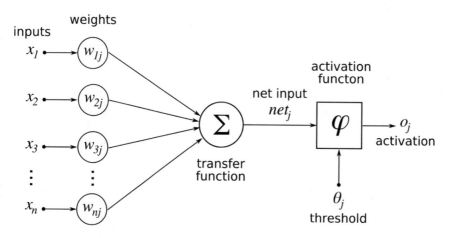

Fig. 8.1 Artificial neuron model

Fig. 8.2 Simplified ANN used with several inputs, hidden neurons with coefficient arrays (W and bias b), and outputs

8.3.2 ANN Energy Consumption Model

Designing ANN based models follows a number of systemic procedures. In general, there are five basics steps: (1) collecting data, (2) pre-processing data, (3) building the network, (4) train, and (5) test the model performance. After data collection, the available data has been pre-processed to train the ANN more efficiently.

ANNs models in this research used the following inputs: electric appliance, as well as apartment area and occupants for a total of 16 inputs. The trained model used the daily electricity consumption recorded data and inputs from a 93 household "training dataset". The number of people of each household has been selected relied with the apartment area as an example of the implication of this method.

This study creates a comprehensive residential energy consumption model using the ANN approach from an energy consumption database consisting of a set of 93 households, recorded in Lisbon (Portugal). Each household have a logged data around 6–8 week, including weekends. Since, ANNs are "data-driven" methods, they typically require large samples during the training stage and they must be appropriately selected. For this study, it was used a complete logged data acquired during six complete weeks at each household for training and validation. A total 93,744 records have been used (24 h × 7 days × 6 weeks × 93 households = 93,744 h).

The mentioned collected data has been pre-processed before being ready to use during the ANN training and testing stages. It has been subjected to some form of preprocessing, that intends to enhance the forecasting model performance, such as "cleaning" the data (by removing outliers, missing values or any irregularities), since during the learning stage the ANN is sensitive to such defective data, degrading the performance of the model obtained.

The missing data are replaced by the preceding or subsequent neighboring values during the same week. Normalization procedure before presenting the input data to the network is generally a good practice, since mixing variables with large magnitudes and small magnitudes will confuse the learning algorithm on the importance of each variable and may force it to finally reject the variable with the smaller magnitude [23].

At the stage of building the network, the researchers specifies the number of hidden layers, neurons in each layer, transfer function in each layer, training function, weight/bias learning function, and performance function. As it is outlined in Sect. 8.4, six different types of training algorithms are investigated for developing the MLP network. During the training network process, the weights are adjusted in order to make the actual outputs (predicated) close to the target (measured) outputs of the network.

This process begun with a network which had 3 neurons in its hidden layer, and repeated, increasing the number of neurons up to 30. The LM algorithm with 20 neurons in the hidden layer for network has produced the best results, and it is used for generating the outputs.

A three-layered feedforward neural network trained by the Levenberg-Marquardt algorithm has been adopted to deliver daily (average and maximum) and hourly energy forecast (one-step ahead forecasts: forecasts for next day's total load; forecasts for hourly loads and define de load profile from a random day). The inputs to train the ANNs proposed in this paper are historical data load, electric appliance, occupancy and area apartment, which will be detailed below. At the training stage, several numbers of neurons in the hidden layer were tested. The best results were produced with twenty hidden neurons. The output layer has one neuron, which was set up to output load forecasts.

In an ANN the relation between inputs and outputs is driven from the data themselves, through a process of training that consists in the modification of the weights associated to the connections, using learning algorithms. In the learning process a neural network builds an input–output mapping, adjusting the weights and biases at each iteration based on the minimization of some error measure between the output produced and the desired output. Thus, learning entails an optimization process. The error minimization process is repeated until an acceptable criterion for convergence is reached.

Using the backpropagation learning algorithm, the input is passed layer through layer until the final output is calculated, and it is compared to the real output to find the error. The error is then propagated back to the input adjusting the weights and biases in each layer. The standard backpropagation learning algorithm is a steepest descent algorithm that minimizes the sum of square errors.

The standard backpropagation learning algorithm is not efficient numerically and tends to converge slowly. In order to accelerate the learning process, two parameters of the backpropagation algorithm can be adjusted: the learning rate and the momentum. The learning rate is the proportion of error gradient by which the weights should be adjusted. Larger values can give a faster convergence to the minimum but also may produce oscillation around the minimum. The momentum determines the proportion of the change of past weights that should be used in the calculation of the new weights.

To avoid overtraining and overfitting it was used the early stopping method [24] to stop the training process. The generated data set was divided into three subsets: a training set, a validation set, and a test set. A training set is a portion of a data set used to train an ANN for the forecasting. The error on the validation set is monitored during the training process and the last set (test set) of parameters was computed to produce the forecasts.

The validation error will normally decrease during the initial phase of training, as does the training set error. However, when the network begins to overfit the data, the error on the validation set will typically begin to rise. When the validation error increases for a specified number of iterations, the training is stopped, and the weights and biases at the minimum of the validation error are returned. The test set error is not used during the training, but it is used as a more objective measure of the performance of various models that have been fitted to the training data.

The evaluation of the ANN performance is based on the shape of the load profile distribution and following criterion: mean absolute percent error (MAPE); the standard deviation of error (SDE) and; serial correlation (linear correlation R and linear regression R^2), are defined as below.

The MAPE criterion is given by:

$$MAPE = \frac{100}{N} \sum_{j=1}^{N} \frac{|p_j - a_j|}{\bar{p}} \tag{8.2}$$

$$\bar{p} = \frac{1}{N} \sum_{j=1}^{N} p_j \tag{8.3}$$

Linear correlation coefficient is given by:

$$R = \frac{N \sum_{j=1}^{N} (p_j a_j) - \sum_{j=1}^{N} p_j \sum_{j=1}^{N} a_j}{\sqrt{\left[N \sum_{j=1}^{N} p_j^2 - \left(\sum_{j=1}^{N} p_j \right)^2 \right] - \left[N \sum_{j=1}^{N} a_j^2 - \left(\sum_{j=1}^{N} a_j \right)^2 \right]}} \tag{8.4}$$

Linear regression is given by:

$$R^2 = 1 - \frac{\sum_{j=1}^{N} (p_j - a_j)^2}{\sum_{j=1}^{N} (a_j)^2} \tag{8.5}$$

where p_j and a_j are respectively the forecasted and record load at hour j, $\bar{p}$ is the average price of the forecasting period and N is the number of forecasted hours.

The SDE criterion is given by:

$$SDE = \sqrt{\frac{1}{N} \sum_{j=1}^{N} (e_j - \bar{e})^2} \tag{8.6}$$

$$e_j = p_j - a_j \tag{8.7}$$

$$\bar{e} = \frac{1}{N} \sum_{j=1}^{N} e_j \tag{8.8}$$

where e_j is the forecast error at hour j and $\bar{e}$ is the average error of the forecasting period.

Load consumption can rise to tens or even hundreds of times of its normal value at particular hours. It may drop to zero or near at other hours. Hence, the average load was used in Eq. (8.2) to avoid the problem caused by load to zero. In linear correlation (R) and regression (R^2) are statistical measure of how well the correlation or regression line approximates the real data points. An R or R^2 of 1 indicates that the correlation or regression line perfectly fits the data.

The next step is test the performance of the developed model. At this stage unseen data are exposed to the model.

8.3.3 Daily Energy Consumption: Average and Maximum

In order to forecast daily energy consumption average and maximum, the inputs of the network were the daily average energy consumption of each electric appliance (lighting, refrigerator, chest freezer, cooking, dishwasher, washing machine, domestic hot water, cooling and heating systems, TV, VCR/DVD, computers and electronic entertainment) the apartment's areas and the occupants of a training households set (47 households).

To achieve the mentioned model the ANN approach use 16 neurons in the input layer, 20 neurons in the hidden layer and 1 neuron in the output layer (Fig. 8.3). The variations in upper number of hidden neurons did not significantly affect forecasting accuracy [13, 25].

For the daily consumption forecast was used a total of 1581 data, 50% of those data, making up the first subset, were used for ANN training and validation. The last 50% of the original data, making up the second subset, was used to evaluate the prediction capacity of the developed ANN model. The training data from the first subset was used for computing the ANN weights and biases, and the validation, from the second subset, was used to test the accuracy of the ANN model.

The first step for ANN training was take 85% of the data from first subset mentioned above for ANN training and validation. During training, the last 15% was used to evaluate the forecast capacity of the developed ANN model.

Hyperbolic tangent sigmoid function (HTSF) and pure line function (PF) were used as the transfer function in the hidden layer and output layer, respectively, and for the input layer it used PF. All data in input and output layers were normalized in the [0, 1] and weights [−1, 1]. In the selected ANN model, inputs were the hourly mean energy consumption of all electric appliance of those direct measurements recorded in those years.

The model was built, trained and tested using several networks each, putting through trial and error a different number of neurons (up to 30) in their hidden layer, using software tool (*nntool*) developed by Matlab. For each trial, the forecasting correlation and error values of the outputs were calculated and compared with the test set (logged data), as well the shape of the load profile distribution.

8.3.4 Hourly Energy Consumption

In order to forecast the hourly energy consumption for a usual day it was used a network with three layers, which is shown in Fig. 8.4. The inputs of the network were the hourly energy consumption of each household, including both weekdays and weekend and the output was the next hourly energy consumption.

An appropriate neural network structure with an adequate initialization method and efficient data processing have been developed to deal and improve nonlinear function and supervised backpropagation algorithm is used to train the networks.

To reduce the complexity of the training task, for the hourly consumption, a total of 3 week data, i.e., 32,760 data (504 h × 65 households = 32,760 h) were used. For each household it was used 504 h, 2/3 of those data (336 h), the two first weeks, making up the first subset, were used for ANN training and validation. The last 1/3 of the original data (168 h), the last week, making up the second subset, was used to evaluate the prediction capacity of the developed ANN model.

The ANN model used Boolean input as an hourly meter system [26] for the forecast hourly load [27]. This method uses an input vector composed of Boolean variable and can take any configuration among 2^n different vectors containing Boolean values such as: 0000...0 = 01:00 h; 1000...0 = 02:00 h; 01000... 0 = 03:00 h; 111...1 = 24:00 h. This model uses a concept of 11 data in the input layer: load energy (5 last hour load values and for 6th hour the load 24 h ago) and binary encoding for the time (0 to 24 h) with n = 5; 20 neurons in the hidden layer and; 1 output (next hourly load) in the output layer.

The network training was performed by using LM backpropagation algorithms, HTSF were used as the transfer function in the hidden layer and output layer, and PF in input layer, also performed under software tool (nntool) developed by Matlab. All data in input and output layers were normalized in the [0, 1] and weights [−1, 1]. In the selected ANN model, inputs were the real-time hourly energy consumption measurements recorded in those weeks.

For this network, the forecasting correlation and error values of the outputs were calculated and compared with test set, as well the shape of the load profile distribution. The effectiveness of this approach is confirmed by forecast outputs obtained. In conclusion, the proposed artificial neural network shows a reliable performance for load forecasting.

8.4 Validation

8.4.1 Database

Despite the study model and database, electric appliance load profile has been validated by Almeida et al. [28] and based on information gathered from monitoring carried out by Sidler [1], the database was validated by this research team.

The database is characterized by 93,744 hourly energy consumption data collected and recorded directly in 93 households, 18 months, in Lisbon, Portugal.

The researchers analyzed, for each household, the status and profile of each registry and corrected the flaws (w/ value: 0 Wh), by adding value antecedent or precedent (choosing the higher value), thus ensuring the profile of the load curve and mislead the ANN.

Therefore, since the all database has been analyzed in detail and amended, the research team validates the database, since, in addition to the reasons stated in this paper, it ensures consistency of its quality for this work as one comes to demonstrate.

8.4.2   Selection of the Network Architecture

In order to determine the best approach model to forecast energy consumption for each household, it was select several network architecture with one hidden layer and the learning algorithm that produces the best prediction performance. Since there are two input layer (16 and 11 to forecast daily and hourly energy consumption, respectively) and one output layer, for this stage, chosen de neural network architecture, the network sample has 16 inputs, 1 output and 20 neurons, the networks were trained with six learning algorithms available in the Matlab software, available at that time de research forwarded.

Thus, a total of 12 networks (6 learning algorithms × 2 network configurations = 12) were tested. The parameters of the learning algorithms used in the analysis are given in Table 8.1. Training was halted when the testing set sum of squares error (SSE) value stopped decreasing and started to increase—an indication of over-training and de MSE get the best low value—and indication of lower square error. The prediction performance in terms of SSE, R^2 and MSE of the networks with the lowest testing set SSE and MSE values amongst the 2 networks for each of the six learning algorithms is presented in Table 8.2.

As can be seen from Table 8.2, the feedforward Levenberg-Marquardt produced the best forecast with the higher R^2 being 0.99962 and faster than the other network with 11 iterations, indicating that this network has the highest performance amongst the networks tested.

By doing this choice, the model structure selection is reduced to dealing with the chosen network inputs/outputs and the internal network architecture. The ANN nonlinear models, expressed by Eq. 8.9, assume that the dynamic process output can be described, at a discrete time instances, by a nonlinear difference equation, depending from the past inputs and outputs [29]:

$$\hat{y}(t) = \hat{f}(y(t-1), \ldots, (t-n), u(t-1), \ldots, u(t-m)) \qquad (8.9)$$

where n and m are the output and input signal delays, respectively, and $\hat{f}$ is a nonlinear function (the MLP activation function). Essentially, the MLP networks are used to approximate the $\hat{f}$ function if the inputs of the network (ui(t)) are chosen as the n past outputs and the m past inputs:

Table 8.1 Parameters of the learning algorithms used

Learning algorithm	Parameters[a]			
Elman backpropagation	μ: 0.001	μ_dec 0.1	μ_inc 10	
Layer recurrent	μ: 0.001	μ_dec 0.1	μ_inc 10	
NARX	μ: 0.001	μ_dec 0.1	μ_inc 10	
Cascade-forward back-propagation	μ: 0.001	μ_dec 0.1	μ_inc 10	
Levenberg-Marquardt	μ: 0.001	μ_dec 0.1	μ_inc 10	
Resilient propagation	β inc.: 1.2	β dec 0.5	$\phi_{initial}$: 0.07	ϕ_{max}: 50

[a]the definitions for the parameters of the learning algorithms are given in the appendix

Table 8.2 The performance of the network trained using six different learning algorithms

Network	Learning algorithm	No of hidden units	No of neurons	No interaction	SSE	R^2	No interaction	MSE	R^2
1	Elman back-propagation	1	20	14	596.000	–	91	215	–
2	Layer recurrent	1	20	24	403.000	–	108	143	–
3	NARX	1	20	27	417.000	–	29	197	–
4	Cascade-forward back-propagation	1	20	54	159.000	0.90515	12	697	0.97721
5	Feed-forward back-propagation								
5.1	**Levenberg-Marquardt**	**1**	**20**	**167**	**164.000**	**0.99817**	**11**	**28**	**0.99962**
5.2	Resilient propagation	1	20	27	66.400.000	0.98709	89	18700	0.99796

$$\hat{y}_i(t) = F_i \left[\sum_{j=1}^{N_h} v_{ij} f_i \left(\sum_{k=1}^{n+m} w_{jk} x_k(t) + b_{j0} \right) + b_{i0} \right], \quad i = 1, \ldots M \quad (8.10)$$

where N_h is the number of hidden layer neurons [$N_h = 16$ (Fig. 8.3) or 11 (Fig. 8.4)], w_{jk} are the first layer interconnection weights (the first subscript index referring to the neuron, and the second to the input), v_{ij} are the second layer interconnection weights, b_{j0} and b_{i0} are the threshold or sets (biases), and F_i and f_i are the activation functions of the output neurons and hidden neurons, respectively.

Thus, the modeling process is to find the neural network that fits this ANN model equation. To obtain the ANN models some steps must been made:

1. The data collection from the process that are representative of a great part of the process working points;
2. The model structure selection;
3. The model estimation;
4. The model validation.

One of the attractive features of using MLP networks to model unknown non-linear processes is that a separate discretization process is rendered unnecessary since a nonlinear discrete model is available immediately [30].

The MATLAB neural network toolbox was used to train the feedforward backpropagation ANN models, because it had structure selection, estimation, validation and simulation (forecast) in a GUI box. MATLAB provides built-in transfer

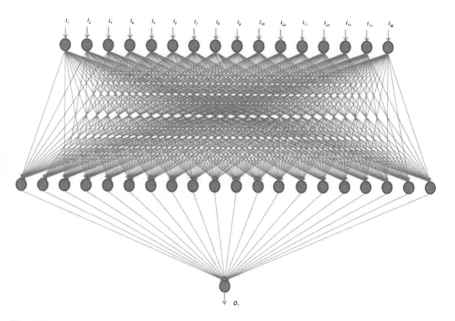

Fig. 8.3 ANN architecture used for daily energy average and maximum consumption

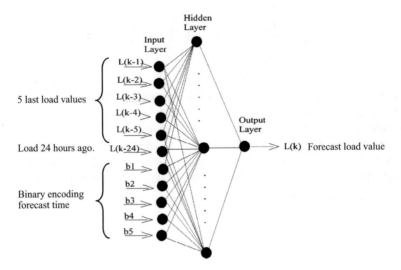

Fig. 8.4 Simplified ANN Architecture used to forecast hourly load

functions which are used in this study: linear (purelin) and Hyperbolic Tangent Sigmoid (HTSF); for the hidden layer and linear functions for the output layer. This function is a nonlinear, continua and differentiable function, as needed for back-propagation training algorithms. The graphical illustration and mathematical form of such functions are shown following:

Linear:

$$\zeta = b + \sum_{k=1}^{n} w_k x_k \tag{8.11}$$

Hyperbolic Tangent Sigmoid:

$$\zeta = \tanh\left(b + \sum_{k=1}^{n} w_k x_k\right) \tag{8.12}$$

After this, the determination of the number of hidden neurons has been made by an iterative process. The training process was based on the MSE (Mean square error), the number of neurons in the hidden layer was iteratively reduced to preserve the model performance and, to be the less possible, to ensure generalization capabilities and avoid overfitting.

The algorithm that has been chosen for train the ANN networks models has been the Levenberg-Marquardt algorithm because is the fastest algorithm, and with better results, on function approximation problems as shown in [31]. In general, on function approximation problems, for networks containing up to a few hundred weights, the Levenberg-Marquardt algorithm will have the fastest convergence (but, it needs much more memory).

The neurons activation functions used are the hyperbolic tangent for the hidden layers and linear and hyperbolic tangent function for the output layers, because it is necessary that the ANN weights should be between $[-1, 1]$.

During training the ANN, despite requiring more memory and more time spent for calculation, it was found that hyperbolic tangent activation function provides better performance and lower error (MSE). Thus, it was decided to present the work results with the hyperbolic tangent function.

The feedforward Levenberg-Marquardt algorithm with 20 neurons produced the best forecast with the higher R^2 being 1 and lower MSE with 3.53, indicating that this architecture has the highest performance amongst the architecture tested.

Han et al. [32] used the predicting ANN model 20 neuron that are chosen to match the average number of hidden neuron in the adaptive network.

In futures works, to choose the number of hidden neurons in neural networks, for identification problems, Mendes et al. [31] propose a new approach process of choosing the number of hidden neurons using a pruning algorithm.

8.5   Results and Discussion

The ANN model was trained using the database energy consumption from 93 households, values from years 2000–2001 (18 months), and the trained model was then tested and compared with another set of daily and hourly energy consumption. For learning the neural network, that was adopted the well-known Levenberg–Marquardt back-propagation algorithm and activation function used was the hyperbolic tangent for both hidden and output layers. Also, the training process used the LM algorithm with 20 neurons, because produced the best forecast with the higher R^2 and lower MAPE and SDE. The training of neural network continues unless and until the error becomes constant. After the error becomes constant, the learning procedure terminates.

It started working with the daily average consumption of each household, and concludes with hourly energy consumption. Training models as ANN leads to the conclusion that they were non-linear functions. It found that the proposed artificial neural network (ANN) has an advantage of dealing not only with non-linear part of the load, was able to forecast the daily and hourly energy consumption with good result and precision for a normal day, weekend and special days.

It was tried to find other results with this research, as the neural network in forecasting the maximum consumption for each household, Boolean input application as an hourly meters system and compare the results forecast with real values measured and logged in the database mentioned.

The daily, maximum and hourly consumption has an important role in shaping the energy storage design. Knowing the forecasting consumption, power production and hourly energy demand is possible shedding (anticipate or postpone) the consumption of electricity.

8.5.1 Daily Energy Consumption: Average and Maximum

The neural network model was trained by using data from a random day of each household and tested to forecast a typical day to another household. For each those clusters, to predicted household's daily energy consumption, the results were good. This can be attributed to the amount of daily and hourly training data used to tuning the model [33].

In order to forecast maximum power consumption in a household, it was collected the highest data energy consumption of each day and household. It can be concluded, as the latter work for the daily energy consumption for a random day, the output of ANN with 20 neurons are very close to the real values.

By analyzing the Table 8.3, Figs. 8.5 and 8.6, it is shown that the adopted model for predicting average and maximum consumption of housing allows good outcomes. Correlation values and linear regressions near unity validate the performance and adjustment of results. The values of the error (MAPE) and standard deviation (SDE) confirm sureness in the model adopted. These forecasting are reliable and satisfactory for power demanding and energy storage for each household.

The figures show the modeling results of Portuguese household daily and maximum energy consumption (in the bottom, the horizontal axis, each number identified one household and vertical axis the daily consumption in kWh/day). The blue symbol ($*$) represents the daily electric energy consumption, measured and recorded in a random day, and the red box ($\Box$) line represents the ANN forecast for the same day.

8.5.2 Hourly Energy Demand

Simple Forecast ANN Model

Using the same ANN daily methodology, the following figures show the performance of the neural network to a randomly selected household and it appears that is satisfactory and interesting ANN behaves.

The profiles shown in Figs. 8.7, 8.8, 8.9, 8.10, 8.11, 8.12 and 8.13, allow us to evaluate that are good results early hours forecasting. As one moves away from a week, the performance decreases and the forecast error increased as expected.

Table 8.3 Statistical analysis of the forecasting error attained with the neural network approach for 46 households analyzed

Daily energy	R (%)	R^2 (%)	MAPE (%)	SDE
Average	98.9	97.9	4.2	1.06
Maximum	93.5	87.4	18.1	8.25

Fig. 8.5 Comparison of daily energy demand [kWh] by household

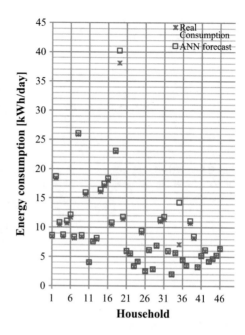

Fig. 8.6 Comparison of maximum consumption [kWh] for each household

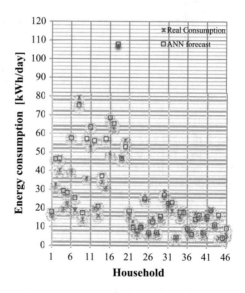

Boolean Input Forecast ANN Model

The ANN model used the Boolean input application, with 11 inputs, one hidden and one output layer, as shown in Fig. 8.4, and for learning the neural network, that was adopted the Levenberg–Marquardt algorithm and a network which had 20 neurons in its hidden layer.

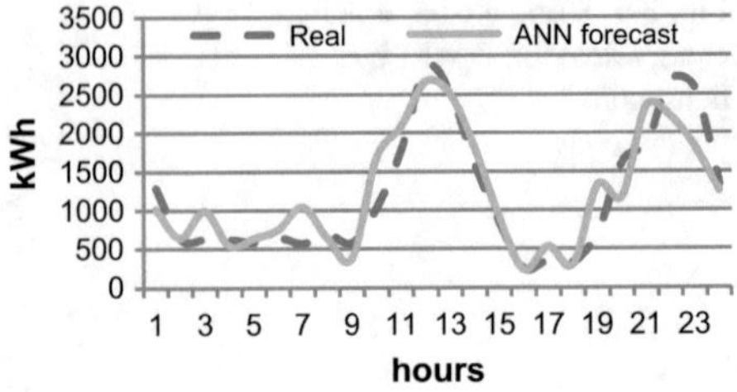

Fig. 8.7 ANN—Comparison
of hourly consumption—1st
day (Saturday)

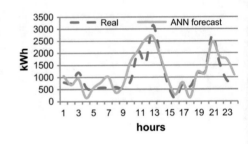

Fig. 8.8 ANN—Comparison
of hourly consumption—2nd
day (Sunday)

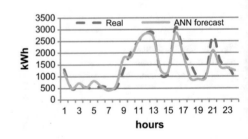

Fig. 8.9 ANN—Comparison
of hourly consumption—3rd
day (Monday)

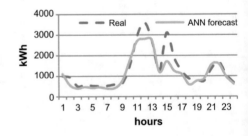

Fig. 8.10 ANN—
Comparison of hourly
consumption—4th day
(Tuesday)

The ANN was trained using the first two weeks of data, one random household
and tested for the next 3 day of 3rd week. Figures 8.14 and 8.15 show the forecast
hourly load of random households. Evaluating those Figures and Table 8.4 it is
found that for daily profiles which have a similar pattern the method proposed by
[27, 34] has a good performance for the first three days.

In test process, the forecasting correlation and error values of the outputs were
calculated are:

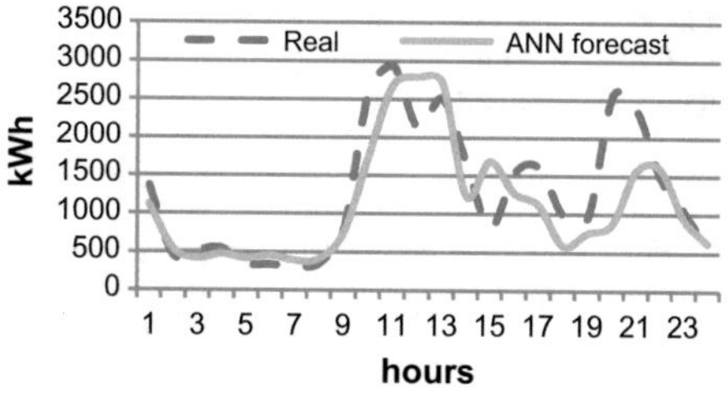

Fig. 8.11 ANN—Comparison of hourly consumption—5th day (Wednesday)

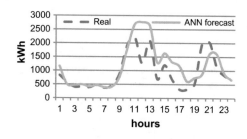

Fig. 8.12 ANN—Comparison of hourly consumption—6th day (Thursday)

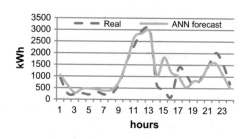

Fig. 8.13 ANN—Comparison of hourly consumption—7th day (Friday)

Fig. 8.14 Forecast hourly load of household no 64—first 3 days of 3rd week

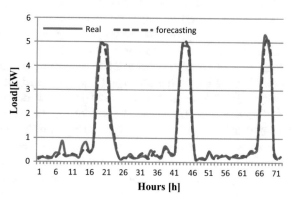

Fig. 8.15 Forecast hourly
load of household no 65—
first 3 days of 3rd week

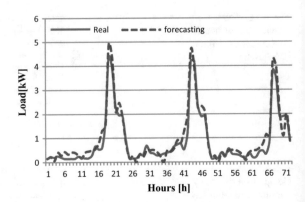

Table 8.4 Statistical analysis of the forecasting error attained with the neural network approach
for the households H64 (left) and H65 (right) analysed

	R (%)	R^2 (%)	MAPE (%)	SDE	R (%)	R^2 (%)	MAPE (%)	SDE
1st	98.8	97.7	16.0	0.23	98.3%	96.6%	21.5	0.17
2nd	99.6	99.3	10.0	0.09	97.6%	94.5%	19.3	0.21
3rd	99.3	98.6	12.9	0.14	97.9%	93.2%	23.5	0.22

8.5.3 Results Obtained in Similar Studies

The analysis of the current literature on the use of MLP neural networks with
backpropagation forecasting a daily load curve with an advance of 24 h, reveals a
small number of papers mostly dedicated to large power systems operation as
electric companies. These methods are able to provide good results and the use of
temperature ensures an improved quality of forecasts [35–38].

However, comparing the work shown in this paper with the existing works in the
scientific literature, the approach followed in this work forecasts electric load for
long periods ahead (i.e., over 24 h), temperature and other weather data are not
considered in these predictions and it was not used the group of days (weekday and
weekend). The neural network used in this work is fed with actual load values
available in order to forecast the next period. In [34] a 30 days forecast (following
month) using data of 30 days (preceding month) is shown. The present approach
shows a 3 days forecast using 15 days (two weeks) data with an ANN that is
simpler providing results that present the same level of accuracy.

The figures show the modeling results of Portuguese household hourly energy
consumption (in the bottom, the horizontal axis, each number identified one hour
and vertical axis the energy consumption in kWh). The straight blue line represents
the hourly load, measured and recorded in the first 3 days of the 3rd week, and the
discontinuous red line represents the ANN forecast for the same days.

8.6 Conclusion

As current literature, these results show that the ANN technique can be reliably used in forecasting daily and hourly electric energy consumption and evaluated the performance of the neural network for forecasting the hourly energy demand. It will adjust the network and make the forecast of precedent hourly consumption for each household.

This paper introduced forecasting methods of daily and hourly energy consumption, by using an Artificial Neural Network (ANN). After identifying a faster algorithm, such as Levenberg-Marquardt, the research reveal that ANN are able to forecast daily and hourly energy consumption, as well the load profile with accuracy. This can be attributed to the amount of daily and hourly training data used to tuning the model [33].

Cluster analysis method has been applied based on electric consumption and occupancy patterns. To determinate the electric load profile, the required input data are hourly and daily data measured end-use energy consumptions. The daily energy consumption and electric load profiles have been determined using several weeks, including both weekdays and weekend.

The paper introduced the ANN technique for modeling energy consumption for a random day, the next hours (until 3rd day), using a Boolean metering system. To determinate the electric load profile, the required input data are hourly and daily data logged end-use energy consumptions. The energy consumption and electric load profiles have been determined using several weeks, including both weekdays and weekend.

The apartment area and inhabitants inputs are the two variables it define the shape of the forecasting. The input data as kitchen appliances, lighting, cooling and heating, domestic hot water (DHW) and entertainment appliances have negligible contribution to the improvement of the forecasting performances. This analysis allows concluding that, in the region of Lisbon, the households are provided with similar electrical appliances. However, this extrapolation to other regions and countries should be carefully evaluated.

Forecast hourly and daily energy consumption can be useful to determine the required size of a storage energy system, delay and postpone energy consumption, and can be used at the renewable energy system early design stage. It can also help on the demand-side management (DSM), such as electricity suppliers, to forecast the likely future development of electricity demand in the entire sector of the community.

For future research, an important step is continuing this additional research enhancing forecasting capability on the load profile renewable energy production (micro production) to identify an accurate, effective and appropriate renewable energy production and energy storage.

Acknowledgements This work was partially supported by FCT, through IDMEC, under LAETA PestOE/EME/LA0022 and by the Sustainable Urban Energy System (SUES) project under the MIT Portugal Program (MPP).

References

1. Sidler, O.: Demand-side management end-use metering campaign in the residential sector, European community SAVE programme (2002)
2. Hippert, H.S., Pedreira, C.E., Souza, C.R.: Neural networks for short-term load forecasting: a review and evaluation. IEEE Trans. Power Syst. **16**(1), 44–55 (2001)
3. Carpinteiro, O., Reis, A., Silva, A.: A hierarchical neural model in short-term load forecasting. Appl. Soft Comput. **4**, 405–412 (2004)
4. Gross, G., Galiana, F.D.: Short-term load forecasting. Proc. IEEE **75**(12), 1558–1573 (1987)
5. Taylor, J.W.: Short-term load forecasting with exponentially weighted methods. IEEE Trans. Power Syst. **27**, 458–646 (2012)
6. Feinberg, E.A., Genethliou, D.: Load forecasting. Applied Mathematics for Restructured Electric Power Systems: Optimization Control, and Computational Intelligence, vol. Chapter 12 (2005)
7. AlFuhaid, A.S., El-Sayed, M.A., Mahmoud, M.S.: Cascaded artificial neural networks for short-term load forecasting. IEEE Trans. Power Syst. **12**(4), 1524–1529 (1997)
8. Aydinalp, M., Ugursal, V., Fung, A.: Modeling of the appliance, lighting, and space cooling energy consumptions in the residential sector using neural networks. Appl. Energy **72**(2), 87–110 (2002)
9. Beccali, M., Cellura, M., Lo Brano, V., Marvuglia, A.: Forecasting daily urban electric load profiles using artificial neural networks. Energy Convers. Manage. **45**, 2879–2900 (2004)
10. Beccali, M., Cellura, M., Lo Brano, V., Marvuglia, A.: Short-term prediction of household electricity consumption: Assessing weather sensitivity in a Mediterranean area. Renew. Sustain. Energy Rev. **12**, 2040–2065 (2008)
11. Tso, G., Yau, K.: Predicting electricity energy consumption: a comparison of regression analysis, decision tree and neural networks. Energy **28**, 1761–1768 (2005)
12. Aydinalp, M., Ugursal, V.: Comparison of neural network, conditional demand analysis, and engineering approaches for modeling end-use energy consumption in the residential sector. Appl. Energy **85**, 271–296 (2008)
13. Khotanzad, A., Afkhami-Rohani, R., Abaye, A., Davis, M., Maratukulam, D.: ANNSTLF—a neural-network-based electric load forecasting system. IEEE Trans. Neural Networks **8**(4), 835–846 (1997)
14. Kalogirou, S.A.: Applications of artificial neural networks in energy systems: a review. Energy Convers. Manage. **1999**(40), 1073–1087 (1999)
15. Kalogirou, S.A.: Applications of artificial neural-networks for energy systems. Appl. Energy **67**(1–2), 17–35 (2000)
16. Kalogirou, S.A.: Artificial neural networks energy systems applications. Renew. Sustain. Energy Rev. **5**, 373–401 (2001)
17. Sietsma, J., Dow, R.: Neural net pruning: why and how. In: Proceedings of IEEE International Conference on Neural Networks, vol. 1, pp. 325–333 (1988)
18. Saini, L.M., Soni, M.K.: Artificial neural network based peak load forecasting using Levenberg-Marquardt and quasi-Newton methods. IEE Proc. Gen. Transm. Distrib. **149**(5), 578–584 (2002)
19. McCulloch, W., Pitts, W.: A logical calculus of ideas immanent in nervous activity. Bull. Math. Biophys. **5**(4), 115–133 (1943)
20. Hebb, D.: The Organization of Behavior (1949)
21. Farley, B., Clark, W.A.: Simulation of self-organizing systems by digital computer. IRE Trans. Inf. Theor. **4**(4), 76–84 (1954)
22. Rochester, N., Holland, J.H., Habit, L.H., Duda, W.L.: Tests on a cell assembly theory of the action of the brain, using a large digital computer. IRE Trans. Inf. Theor. **2**(3), 80–93 (1956)
23. Tymvios, F., Michaelides, S., Skouteli, C.: Estimation of surface solar radiation with artificial neural networks. In: Badescu, V. (ed.) Modeling Solar Radiation at the Earth's Surface: Recent Advances, p. 221–256 (2008)

24. Sarle, W.S.: Stopped training and other remedies for overfitting. In: Proceedings of the 27th Symposium on the Interface of Computing Science and Statistic, pp. 362–360 (1995)
25. Bakirtzis, A.G., Petridis, V., Kiartzis, S.J., Alexiadis, M.C.: A neural network short term load forecasting model for the Greek power system. IEEE Trans. Power Syst. **11**(2), 858–863 (1996)
26. W. Holderbaum, Canart, R., Borne, P.: Artificial neural networks application to boolean input systems control. Laboratoire d'Automatique et d'Informatique Industrielle de Lille (1999)
27. Lacerda, W.S.: Conference at Universidade Federal de Lavras (UFLA), Nov 2006
28. Almeida, A., Fonseca, P., Schlomann, B., Feilberg, N., Ferreira, C.: Residential monitoring to decrease energy use and carbon emissions in Europe. Presented in international energy efficiency in domestic appliances and lighting conference 2006, 2006
29. Mendes, M.J.G.C., Calado, J.M.F., Sá da Costa, J.M.G.: Multi-agent approach to fault tolerant control systems. Tese de Doutoramento em Engenharia Mecânica (em inglês) (2008)
30. Norgaard, M., Ravn, M.O., Poulsen, N.K., Hansen, L.K.: Neural Networks for Modelling and Control of Dynamic Systems. Springer (2000)
31. Mendes, J.G.C., Calado, J.M.F., Sá da Costa, J.M.G.: Pruning algorithm applied to a hierarchical structure of fuzzy neural networks: case study. In: 8th IEEE International Conference on Methods and Models in Automation and Robotics—MMAR 2002, vol. 1, p. 207–212. 2–5 Sept 2002
32. Han, H.G., Wu, X.L., Qiao, J.F.: Real-time model predictive control using a self-organizing neural network. IEEE Trans. Neural Networks Learn. Syst. **24**(9) 1425–1436 (2013)
33. Mikhalakakou, G., Santamouris, M., Tsangrassoulis, A.: On the energy consumption in residential buildings. Energy Buildings **34**(7), 727–736 (2001)
34. Zebulum, R.S., Vellasco, M., Guedes, K., Pacheco, M.A.: Short-term load forecasting using neural nets. In: Proceedings Natural to Artificial Neural Computation, Springer, Berlin—International Workshop on Artificial Neural Networks Malaga-Torremolinos, vol. 930, ISBN 978-3-540-49288-7, pp. 1001–1008. 7–9 June 1995
35. Afkhami, R., Yazdi, F.M.: Application of neural networks for short-term load forecasting. In: IEEE Power India Conference, 10–12 Apr 2006
36. He, Y.-J., Zhu, Y.-C., Gu, J.-C., Yin, C.-Q.: Similar day selecting based neural network model and its application in short-term load forecasting. In: Proceedings of International Conference on Machine Learning and Cybernetics, vol. 8, pp. 4760–4763. 18–21 Aug 2005
37. Ramezani, M., Falaghi, H., Haghifam, M.R., Shahryari, G.A.: Short-term electric load forecasting using neural networks. In: The International Conference on Computer as a Tool—EUROCON 2005, vol. 2, pp. 1525–1528. 21–24 Nov 2005
38. Barghinia, S., Ansarimehr, P., Habibi, H., Vafadar, N.: Short term load forecasting of Iran national power system using artificial neural network. In: Power Tech Proceedings, 2001 IEEE Porto, vol. 3, 10–13 Sept 2001

Chapter 9
Development of a Holistic Method to Analyse the Consumption of Energy and Technical Services in Manufacturing Facilities

John Cosgrove, John Littlewood and Paul Wilgeroth

Abstract This chapter describes the background to energy usage in production operations and sets out some principles, process steps and methods to provide a more holistic view of the Significant Energy Users (SEUs) and the related consumption of energy and technical services (Heat, Air, Water). A model based on direct and indirect energy analysis from a 'product' viewpoint is extended to identify waste or auxiliary energy in line with 'Lean' principles. The auxiliary energy identified represents the best opportunity to gain energy savings through operational and behavioral changes at the lowest possible cost. The proposed process mapping methodology [Value Stream Mapping (VSM)] effectively acquires production and energy data that can be modelled to provide both steady state and dynamic energy consumption and potentially provide a multidimensional hierarchical view of this energy consumption and cost directly related to production equipment. The method is one that can be updated easily to reflect changes in the production environment and to provide a holistic view of the energy and technical services in the context of the varying production activity.

Keywords Energy efficiency in industry · Significant energy users (SEUs) · Technical services · Value stream mapping (VSM)

9.1 Energy in Industry

9.1.1 Introduction

Worldwide, industry consumes [1] almost one-half of all commercial energy used and is responsible for roughly similar shares of greenhouse gases. In the EU28 Countries, the industry sector still accounted [2] for about a quarter of the final

J. Cosgrove (✉) · J. Littlewood · P. Wilgeroth
Cardiff Metropolitan University, Cardiff, UK
e-mail: john.cosgrove@lit.ie; jcosgrove@cardiffmet.ac.uk

© Springer International Publishing AG 2017
J. Littlewood et al. (eds.), *Smart Energy Control Systems for Sustainable Buildings*,
Smart Innovation, Systems and Technologies 67,
DOI 10.1007/978-3-319-52076-6_9

energy consumption in 2012. In absolute terms industrial final energy consumption has decreased from 15.4 million TJ in 1990 to 11.8 million TJ in 2012. This is driven by energy efficiency improvement but also by the slowdown in world production since 2007.

The more industrialised nations naturally have higher percentage of consumption, for example, in Germany, industrial electricity consumption accounts [3] for approximately 46% of national energy usage. Similarly in the UK the DUKES Report [4] shows that in 2012 industry accounted for 31% of the total electricity consumed and 25% of the green house gas (GHG) emissions.

To stay competitive in the 21st Century, manufacturing companies need to include sustainability into their manufacturing optimisation schemes. Sustainable Manufacturing (SM) is the new paradigm [5] for manufacturing companies and involves the integration of all relevant dimensions that affect or have effects on third parties while conducting manufacturing operations, including energy, environmental impact and lifecycle analysis.

Hence, when designing or improving a manufacturing system an alignment with economic, ecological, and social goals has become an essential strategic objective of manufacturing companies [6]. An isolated consideration of traditional economic variables without evaluation of ecological and social impact is no longer acceptable and a balance between traditional material, equipment and personnel resources is required.

To allow for sustainability and to meet growing environmental legislative requirements, industry must be capable of understanding their energy requirements, their energy consumption and the manner in which this is managed, particularly in the production environment. Although there are various sustainability assessment tools available, these tools are complex, require large amounts of data and technical expertise to utilise [7].

Hence, this chapter proposes a practical and less-complex methodology for the assignment of energy usage in technical services (water, HVAC) in Advanced Manufacturing Facilities. This methodology uses a combination of the lean manufacturing principle of Value Stream (VS) Mapping with energy management and on application to a standard manufacturing site, outlines the process flow, energy metering requirements, the technical utilities servicing the process and an identification of significant energy users. This methodology allows a manufacturing company to visualise their production process from an energy perspective and determine the next steps for improvement in energy management and consumption.

In productive industries some consumption of energy is necessary and indeed unavoidable as fundamental principles of energy conversion apply. The key principles [8] of energy reduction in industry are;

1. Avoid unnecessary consumption of energy.
2. Turn things off when they not needed to add-value and ensure that the workforce are engaged with and implement energy efficient operations.
3. Reduce the quantity of energy consumed during the production process
4. Re-design inefficient products and processes and reduce the supply of services in line with production activity.

5. Recover energy from the production process and reuse it.
6. Where waste cannot be eliminated, develop systems to re-use the energy either internally or externally.
7. Generate alternative energy sources (solar energy, wind turbines).

Rahimifard et al. [9] describes how a significant proportion of energy used in manufacturing is currently generated through fossil fuels and that in the foreseeable future, the rationalisation of energy consumption still provides the greatest opportunity for the reduction of greenhouse gases.

While the share of renewable energy production has grown at a greater rate (from 14% in 2004 to 25.4% in 2013) than the growth in overall energy production [2] and even though the adoption of renewable energy technologies may well be the long term solution, increasing the efficient use of energy can generate the greatest and most economical contribution to climate mitigation that industry can deliver.

As set out in the EU Factories of the Future (FoF) roadmap [10], the smart sustainable factory of the future will be one where there is full integration between the production activity and the associated energy used and where the operation of the factory can be optimised around its energy and ecological impact. Thus expected features would include; energy efficient plant and processes that are constrained by the available of energy (time/type), integrated on-site renewable energy generation with export capacity and integrated community based energy sinks/sources to avoid distribution losses and to eliminate waste heat/energy.

9.1.2 Sustainable Manufacturing

Sustainable manufacturing is a term that is commonly used in industry. It is basically the process of lowering the use of energy and utilities while producing the same level or greater of product. The US Department of Commerce's Sustainable Manufacturing Initiative [11] sums it up as:

> The creation of manufactured products that use processes that minimize negative environmental impacts, conserve energy and natural resources, are safe for employees, communities, and consumers and are economically sound.

The OECD [12] has set out a seven step plan to increased levels of sustainable manufacturing in industry, although it is stressed that the seven steps are not necessarily a one way journey. Sustainable manufacturing is not about a final destination or result but rather is about continuous learning, innovation and improvement.

- Step 1: Map impact and set priorities
- Step 2: Understand data needs and choose indicators
- Step 3: Measure inputs used in production
- Step 4: Assess the operations of the facility
- Step 5: Evaluate the product design

- Step 6: Understand the results
- Step 7: Take action to improve performance

Step 1: Map impact and set priorities

It is necessary to understand the overall impact of the factory in terms of environmental impacts from the consumption of electricity, fossil-based fuels, biomass and water. At the highest level possible in the organisation it is necessary to clearly set out the business expectations and the level of priority for energy reduction activities. In larger organisations this commitment is often evidenced through the appointment of an Energy Manager and the creating of appropriate reporting structures to senior management.

Step 2: Understand data needs and choose indicators

It is necessary to gather the data required to achieve your goals. Once the organisation priorities have been identified a data collection process needs to be established. Basically this means that the company must decide what to compare against what and how they are going to measure it. The appropriate metrics [Key Performance Indicators (KPIs)/Energy Performance Indicators (EnPIs)] need to be selected to best represent the company priorities and available data. Both specific energy consumption (Energy Intensity) indicators and absolute energy indicators should be developed.

Step 3: Measure inputs used in production

To clearly understand the drivers of energy consumption in production it is necessary to understand the variation in inputs to production and the units of production output that are measurable in the organisation. It is best to select indicators and a method of normalisation of data that will be consistent across value stream and through changes in seasons, weather, production patterns and production volumes. This ensures that over the period of a number of years of historical data all the years are measured using the same KPIs thus ensuring uniformity and accurate comparison of the data.

Step 4: Assess the operations of the facility

It is necessary to review the existing processes and to determine which operations can be modified in order to reduce energy consumption. Establishing a list of Significant Energy Users (SEUs) and a quantified and prioritised Register of Opportunities is standard practice under most energy management systems.

Step 5: Evaluate the product design

It is necessary for the company to critically evaluate their products and to review if changes can be made to the product design, product assembly steps and/or production processes to see if lower impact can be designed into the facility based on a better understanding of the facility operations. This is easier to achieve in industries with regular change-overs in product design and may not be possible in some regulated industries such as medical devices or pharmaceuticals.

– **Step 6: Understand the results**

After completing the previous steps it is very important to use the figures found as a benchmark for future revision. It is difficult to know if results are good or bad without having something to measure against. This benchmarking the results gives you a powerful tool in the future to check and compare results against. Where possible results should be compared against any available external benchmarks.

Step 7: Take action to improve performance

Once the steps have been set out it is important to develop an action plan to improve efficiency and put a team together to do so. For continued success the action plan must contain concise actions for each member of the team as well as clearly defined timelines. Improvements in energy efficiency are unlikely to be maintained unless there is appropriate energy management tools, metrics and systems in place to provide regular review and control.

9.1.3 Energy Efficiency Potential

Energy Efficiency is the relationship between production output and energy input. Increasing energy efficiency results in increased production for the same amount of energy or a reduction in the amount of energy required for a production unit.

Potential reductions in energy [13] of 20% and in GHG emissions [14] of 30% have been identified on industrial sites where there is an in-depth understanding of energy flows in the manufacturing process and a clear analysis of energy usage. However, to achieve these potential reductions some changes in business practices are needed.

A study [15] conducted by the IEA in 2007 concluded;

> that manufacturing industry can improve its energy efficiency by an impressive 18–26%, while reducing the sector's CO_2 emissions by 19–32%, based on proven technology. Identified improvement options can contribute 7–12% reduction in global energy and process-related CO_2 emissions.

Similarly, a study [16] from 2010 on the energy savings potential in Danish industries found that with a maximum two year payback time requirement, 10% of the final energy consumption could be saved through well proven technologies and energy efficient behaviour. If the payback time requirement is extended to 4 years then the potential savings increased to 15%.

Other studies [17, 18] have shown that where large industries effectively implementing an energy management system (EnMS) that they can reduce their energy use by between 10 and 40%.

In the US, multiple studies [19, 20] have found that industrial facilities could cost effectively reduce their energy use and GHG emissions by 14–22% by 2020. Manufacturers who participated in these energy efficiency programmes also experienced significant additional cost savings [21]. These Non-Energy Benefit (NEBs) included improved production, cleaner environments, improved moral and more reliable operation. Previous research [22] has shown that if NEBs are included, the true value of the energy efficiency projects might be up to 2.5 times higher than if looking at the energy efficiency improvements alone. It is important to attempt to capture any such NEBs and their size to provide better justification and wider implementation of energy efficiency projects.

Despite the efforts made over the last 20 years, the research [23, 24] suggests that there remains an important potential to reduce energy consumption in energy intensive industry by 15–25%. The same research indicates that energy management and behavioral changes can achieve up to half of this remaining energy efficiency potential. Several articles [25, 26] show that only a limited number of companies actually focus on managing energy and that cost-efficient energy efficiency measures are not always implemented, explained by the existence of barriers [27]. The main reasons given for not managing energy is lack of time, lack of resources, lack of knowledge and a primary focus on production.

Seow and Rahimifard [28] conclude that despite the number of commercial tools being used to track and monitor energy use in a factory and across various workstations, the detailed breakdown of energy consumption within various processes and, more importantly, its attribution to total energy required for the manufacture of a unit product is not well understood.

9.1.4 Energy Efficiency Results

Where companies have engaged with energy audit and management scheme, good results have been reported, however the actual results are considerably smaller that the potential outlined above.

In Belgium, over 4000 companies participated, between 2005 and 2014, in a voluntary Benchmarking and Auditing Covenant which is a state run programme on energy efficiency that targets medium sized enterprises. The analysis [29] of the results showed an 8–10% improvement in energy efficiency over the period. Of note, 59% of the energy saving measures stemmed from process-related activities.

In Sweden, a study [30] based on data from 58 industrial firms in the spring of 2012 returned an analysis that they had achieved an energy efficiency improvement of 12% on average. This stemmed from both technology and energy management measures and in absolute numbers provide energy savings of 1,100,000 MWh/year.

In the US, a New York State programme supports Energy Audits of large and medium industries through provision of 50% funding. The audits provide a prioritised list of quantified investment grade energy efficiency measures. Analysis [31] of 303 companies who participated in the programme showed that only 25% of the proposed measures were adopted within the first year. More promisingly, approximately 65% of the proposed measures were installed by the end of the 4th year and there was evidence that the audit report was still being used for guidance more than 6 years after delivery.

9.2 Industrial Energy Classification

9.2.1 Production Versus Facilities

The traditional approach to energy management tends to view a factory in terms of its functional units, i.e. production operations and facilities operations, see Table 9.1. Responsibility for the cost of energy generally lies with the facilities function and thus the greatest focus on energy efficiency is driven by facilities. This is evident in research [32] which highlights the relative priorities of the different functions in industry.

An alternative approach [33] that views energy consumption in terms of how it impacts a product passing through a factory would suggest a different break-down of energy consumption with the costs more fairly assigned to the different functions, as in Table 9.2. A clear decision point is whether that specific consumption of energy would be necessary if there is no active production.

A UNIDO working paper [34] classifies end-use energy consumption in the industrial sector as either Process or Generic. Process refers to energy used directly in the production process, whereas Generic refers to energy used for non-core applications such as heating, ventilation and air conditioning (HVAC), lighting and information technology. However, they state that the boundary between these two categories is not always clear.

As stated, there may be blurred divisions in some areas which in specific factories may warrant a variation on the proposed approach, e.g. vacuum extraction

Table 9.1 Traditional approach to energy classification

Production operations	Facilities operations
Production machines	Electrical utility supply
In-process heating	Heating, ventilation and air conditioning (HVAC)
	Compressed air
	Process cooling/chilled water
	Process/di-ionised water
	Process heating/steam production
	Combined heat and power
	Canteen
	Factory lighting
	Office computing/IT systems

Table 9.2 Product centred approach to energy classification

Production operations	Facilities operations
Process	Generic
Production machines	Electrical utility supply
In-process heating	Heating, ventilation and air conditioning (HVAC)
Compressed air	Combined heat and power
Process cooling/chilled water	Canteen
Process/di-ionised water	Factory lighting
Process heating/steam production	Office computing/IT systems

from production may be integrated with the overall factory ventilation systems or may be separated out as a significant production energy users in the case of a semi-conductor fab. The supply of chilled water may be integral to the facility HVAC and the operation of the CHP in terms of the heat availability may be directly linked to the need for heat in the production process.

The correct assignment of the significant energy users to the respective production operation is the first step in allocating the cost of energy consumed to the production function and to the specific value streams (VS). Management of specific energy consumption at a value-stream level present the opportunity for the value-stream manager and operators to have an input into the efficient operation of their area and to benefits from improvements they make in energy cost reduction.

A detailed review [35] of industrial facilities in Ireland showed that correctly attributing the direct and indirect energy consumption in a large high-volume manufacturing facility gave a split of 57% direct (process) energy consumption and 43% indirect (generic) energy consumption.

In a similar study [36] in the US the analysis states that the majority of energy at larger industrial sites is used by equipment that is associated with the manufacturing process.

9.2.2   Significant Energy Users (SEUs)

Manufacturing facilities, in addition to basic facility end uses, have a number of very specialized energy systems. Some of these—known as cross-cutting technologies [34]—are common to many or most industries while others are processor industry specific. The major cross-cutting technologies include compressed air, steam systems, fan and pump systems, process heating, process cooling, refrigeration, and specialized process space conditioning. Because of their prevalence and energy intensity these end-use systems have been the subject of considerable interest.

Various Sustainable Manufacturing projects are currently been run by medium to large industries to tackle the increasing cost of energy in production and to develop and disseminate best practice in energy management. Each of these projects has identified the critical energy consuming aspects of their industries or what are normally referred as Significant Energy Users (SEUs). These are listed in Table 9.3.

The Industrial Innovation for Energy Efficiency (I2E2) Centre [37] is an Irish government sponsored Technology Centre, established to facilitate research which will have a direct impact on industry. The I2E2 research focus is on energy efficiency improvements in factories, plant, equipment and buildings. Based on feedback from over 30 large industrial sites they have established priority areas for action including compressed air systems, HVAC, heat recovery and production machines.

The EU Intelligent Energy funded Project [38] entitled Sustainable Energy Savings for the European Clothing Industry (SESEC) is driven by the textile industry across Europe. They have identified the non-existence of partial energy meters as a common problem as typically only the main meters from the utility companies are present, i.e. only global energy is measured. They also state that companies are still suffering from historical low interest in energy efficiency mainly due to the low ratio of energy costs/total costs and low internal knowledge on this area.

The EU funded Project entitled SurfEnergy [39]—Path to Energy Efficiency is driven by the Electronics Assembly Industry across Europe. Their members have reported the largest savings in the development of waste heat and energy recovery, improvements to boilers and the use of CHP. They also point to the significant benefits from Energy systems integration and adoption of best practice.

In India, a report [40] on the best practices in energy efficiency adopted by the Indian manufacturing sector, sets out the most common energy efficiency measures and practices adopted between 2007 and 2011 by the automotive, cement and consumer goods manufacturing sectors. The most popular initiative involved the installation of VSDs on motors. Across all the energy saving measures implemented the average payback period was between 1.5 and 1.6 years.

The University of Daytona Industrial Assessment Centre (UD-IAC) is a comprehensive resource [41] for energy efficiency in manufacturing and industry. It contains over 100 assessment recommendations selected from over 900 industrial energy assessments performed by the UD-IAC since 1981. Their goal is to help small and medium-sized industries to reduce costs and stay competitive. Recipients of their assessments report saving an average of about 10% on their energy costs.

As part of the TEMPO Project [35] in Ireland a detailed survey of the energy profile of six large manufacturing sites was undertaken. The study highlighted the exemplar energy efficiency projects completed by the companies between 2011 and 2013. The largest number of projects were undertaken on compressed air, HVAC and facility lighting.

In Canada, the state body Natural Resources Canada—have produced an Energy Savings Toolbox [42] to guide industry on energy efficiency. Besides a focus on the SEUs below they also include a large number of measures in the building envelope. This in not as apparent in other studies and may have greater relevance in Canada due to their harsh winters.

Table 9.3 Critical areas of focus/SEUs

I2E2	SESEC	SurfEnergy	India Industry	UDIAC	Tempo	Canada
Compressed air systems	Compressed air	Conserve compressed air	Compressed air	Compressed air	Compressed air systems	Compressed air
HVAC	HVAC Vacuum	Enclosed hot air dryers		HVAC	HVAC	Heating and boiler plant
Low grade heat recovery	Heat (boiler and gas/fuel supply)		Waste heat recovery	Combined heat and power. Process heating. Steam. Fluid flow	Direct/indirect process heating. Steam production/system	Process heating/heat distribution. hot water service
Manufacturing equipment	Production machines	Process improvements operations and maintenance	Variable speed drives	Motor drives	Direct process machine drives	
	Lighting		Lighting control systems	Lighting	Lighting	Lighting
				Process cooling	Direct process-cooling/refrigeration	Cooling plant/cooling distribution
		Reduce peak demand	Grid voltage regulation	Electrical		Electrical power distribution
		Benchmarking	Energy efficiency awareness programmes			Food areas
	Logistics		Renewable energy			Water service

From this matrix it is possible to draw up a prioritised list of SEUs that industry should pay attention to in order to achieve energy reductions.

Generic Set of SEUs—Priority list

- Heating Systems for facility and provision of hot water/steam for production processes
- Compressed Air systems
- CHP and waste heat recovery
- Provision of chilled water and di-ionised water
- Facility internal and external lighting
- Electrical tariffs, structures and conversion/distribution
- Production machines and conveyors
- Operating Procedures/Employee Behaviour
- Air Conditioning/Vacuum Extraction

9.2.3 Electric Motors

Across many of the SEUs identified above, the primary conversion of electricity into mechanical force (pumps, fans, compressors, machines) is through electric motors and thus manufacturing industry should have a clear strategy on the use of energy efficiency motors, on the correct sizing of motors, on the use of variable speed drives (VSDs) and on the use of advanced control systems to optimise performance.

An International Energy Agency (IEA) report [43] shows that Electric Motor Systems account for between 43 and 46% of all global electricity consumption and there is potential to cost-effectively improve energy efficiency of motor systems by roughly 20–30%, reducing global electricity demand by up to 10%.

In a US study [44], it is estimated that three phase AC induction motors below 500 horsepower will consume 1224 TWh of electricity in 2015, which represents about 30% of the total projected U.S. electricity use. The study shows that these motors are projected to represent 72% of the total electricity consumption of the industry sector.

An analysis [45] at 25 industrial and infrastructure plants in Switzerland has shown that 87.8% of the total electricity consumption of a plant is due to motor systems. This high number may be due to the inclusion of infrastructure plants (Water/Waste Water treatment) where motive power is prevalent. This study also highlighted that in an analysis of 4142 motors in use in these industries, over 50% of them were older than their rated operating life expectancy. On average, these motors were run twice as long as their expected lifetime. Less than 20% of the motors assessed had variable frequency drives to control their load and of a sample of 100 motors that were analysed in further detail, 68% were oversized compared to their load (average load factor below 60%). The key barriers identified in failing to address motor energy efficiency in industry was that industrial users lacked human resources, time, responsibility, technical know-how and financial resources necessary for the implementation of motor systems efficiency projects.

9.3   Production Centered Energy Management

Production processes consume raw materials and transform them into products and wanted or unwanted by-products and use a significant amount of energy to do so. Some of this energy is used for value-added activities embodied into the form and composition of products, while the rest of the energy is wasted in terms of heat losses and emissions. Hence, manufacturing processes generate a significant environmental impact through energy consumption with related resource depletion and GHG emissions [46]. To understand the consumption of energy in a production environment, it is necessary to outline the energy flow within an industrial facility along with the classification of energy usage and its relationship to processes and production outputs.

Imported energy in the form of electricity, gas or solid fuels, for example, coal or peat, along with onsite renewable energy systems provide the primary energy source for a facility. Solid fuels, oil or natural gas are mainly utilised by energy transformation/generation systems such as boilers to generate heat for process and space heating. Electricity (both imported and generated) is used by energy transformation/generation systems mainly to run electric drives to generate mechanical energy. Typical applications include pumps, fans, air conditioning (chill generation, ventilation), and compressors. The energy carriers are the means by which energy moves through the facility, which include compressed air, hot/chilled water, electricity and steam. Energy utilisation systems are the end users of both the electrical and thermal energy. For example, equipment drives and motors use electricity and clean-lines use hot water. The energy drivers are the variables such as weather changes expressed in degree days and variation in production volumes/type which affect energy consumption [47]. The generalised model is shown in Fig. 9.1.

As an example of this energy flow and transformation, for a technical service such as compressed air, electricity (imported and/or generated from Solar/Wind

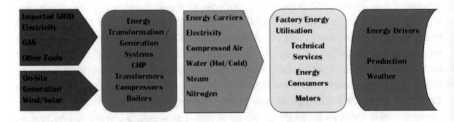

Fig. 9.1 Energy flows in an industrial facility

Turbine) is used by a compressor (energy transformation) to generate compressed air (an energy carrier). The compressed air is used for a product cleaning operation and the main energy driver is production volume, as the use of compressed air will increase as the volume of production increases.

Previous research on manufacturing energy consumption has focused on developing more energy efficient machines/processes [48]. In one case study [9] the effective energy used to directly make the product in a metal processing factory was analysed as 48% of the total energy consumption and is referred to as an efficient production process. In another study [49] the energy requirement for the active removal of material was shown to be quite small compared to the background functions needed for the operation of manufacturing equipment. Drake et al. [50] showed that there are significant amounts of energy associated with machine start-up and machine idling. As a result, in a mass production environment, more than 85% of the energy may be utilised for functions that are not directly related to the production of parts.

This suggests that energy saving efforts, which focus solely on updating individual machines or processes may be missing a significant and perhaps a bigger opportunity. Hence a more holistic mapping of the relationship between production and energy consumption should be applied as research [51] also suggests that a lack of understating between production operations and energy usage prevents energy efficient decision making in real-time. With knowledge of the direct and indirect energy flows and their relationship to production activities it is possible to identify [52] the auxiliary (non-value added) energy within production where there may be significant potential for energy reduction. Developing a clear link between the temporal profile and/or efficiency of the energy consumption by the specific value stream can provide full transparency in the impacts (costs, emissions) of energy consumption and can provide positive feedback and cost reduction to reward improved performance by the value stream.

Currently, the skills and effort needed to identify and to describe the most important machines in a factory and to estimate their energy savings potential is described [45] as considerable—up to 10% of the energy budget before any saving is made. Top level analysis of energy usage, in some cases done purely at the utility bill or at a busbar level only identifies the amount of energy used. A time based energy profile of process level usage is required to support knowledgeable assessments of where improvements can be made throughout an organization while supporting the requirements of the business operations. The ability to identify and determine the impact of Significant Energy Users (SEU's) within an organization with detailed measurement of energy usage in real time can provide an awareness of the profiles and patterns of energy usage throughout a manufacturing or industrial process. Decision support systems and organization management are aided by the knowledge gained when evaluating the energy usage of process structures and components in a manufacturing environment whilst avoiding impact on production. The ability to link Production and Energy models is a vital link in the future application of demand side management to industry.

9.3.1   Monitoring and Targeting

Energy Efficiency begins with measuring energy consumption and continually revisiting those measurements. It is important to know the status of the current consumption profiles before taking improvement actions. However, in order to optimise the energy consumption, besides measuring energy consumption itself, it is important to observe process and activities leading to (higher than necessary) energy consumption. Therefore, it important to identify the most suitable approach to effectively monitor various manufacturing processes and correlate these measurements with the measured energy consumption data to identify what improvements in the processes could lead to reduced energy consumption.

Acquiring and using such context related knowledge is currently very time and cost intensive, for both users and vendors of industrial plants (machines/equipment), especially taking into account the recent trends in industry are demanding more flexibility (re-configurability) from manufacturing systems, leading to more dynamics in the manufacturing operation and consequently more difficulties in monitoring energy use patterns.

This is especially a problem in flexible discrete manufacturing industry where interaction between the processes/production equipment and human operators leading to energy consumption variations is much more difficult to monitor compared to continuous process industry. The gathering of more detailed energy usage data requires a greater deployment of both standard sensing solutions (temperature, current, flow, etc.) and additional sensing solutions such air-line pressures, compressor rpm, vibration, acoustics, etc.

A good Monitoring and Targeting (M&T) approach requires a structured definition of the energy sources mapped to the specific production line or plant. This may be done by process-flowcharting of the manufacturing process and integration of the sensing information on the flowcharts. The integration of the energy and work flow data in real-time gives a clear view of all of the energy paths through the process and gives a high visibility to the relevant energy sources that critically need to be metered. Thus, only the minimum necessary metering, allied with the use of temporary metering and data logging, can be designed to have the least possible impact on production. In fact, a non-invasive approach to metering through clamp-on and external meters is vital in high volume production processes as any downtime required to fit meters is very expensive and generally not acceptable.

9.3.2   Energy Flows in Production Operations

A review of methodologies [47, 53] that categorise energy usage and energy efficiency in industry highlighted that industrial companies still lack appropriate methods to effectively address energy efficiency in a comprehensive and practical manner. This is primarily due to:

- The complexity of production sites that due to business needs, operate more than one production process.
- Production sites may produce various types of products, each with different energy intensity factors.
- Specific energy consumption depends on the production rate and Significant Energy Users (SEUs) are typically viewed in isolation from production operations rather than in conjunction with it (i.e. cycle time and energy usage analysed together to determine process SEUs).
- Comparing different installations (i.e. process equipment, technical services upgrades) using energy efficiency indicators can lead to misleading conclusions, when attempting to take all variables associated with energy efficiency into account.
- The analysis of thermal energy is considerable more complicated in practice than the analysis of electrical usage.

9.3.3 Methodologies

Hermann et al. proposed work that focused on the optimization of the process chain with the objective of securing the best electric energy efficiency. The study proposed a five step approach using a simulation model. These steps include; (1) Analysis of production process chain; (2) Energy analysis of production and its equipment; (3) Energy analysis of technical building services; (4) Load profile and energy cost/energy supply contract analysis; (5) Integrated simulation and evaluation of the production system. However, the work was not extended to an industrial facility or practical application.

Seow and Rahimifard [28] provide a product perspective of energy monitoring and attribute the energy consumed by the product to both the process and the plant. They describes the 'product' viewpoint and the methodology for using energy consumption data at 'plant' and 'process' levels to provide a breakdown of energy used during production. The approach of using a 'product' viewpoint is very much in line with the now standard approach of 'Lean' manufacturing. This approach postulates that Energy consumption in manufacturing can be categorized into Direct Energy (DE) and Indirect Energy (IE), which constitute the embodied energy of a product. DE is the energy required to manufacture a product in a specific process and can be subdivided further into theoretical energy (TE), the energy necessary for actual value creation and auxiliary energy (AE), the energy required by supporting activities for the individual machine/process. Indirect Energy (IE) is defined as the energy necessary to maintain the production environment (lighting, heating/ventilation). This approach has been extended further in Table 9.4 to outline the Energy Management Opportunities (EMOs) that exist at each level.

The proposed energy breakdown also draws particular attention to the auxiliary energy usage, and the potential areas in the factory where energy efficiency

Table 9.4 Industrial energy breakdown

Primary energy		Energy breakdown	Focus areas	EMOs	Potential for savings
Industry primary energy	Direct energy (DE)	Value-added energy (VAE)	Mechanical processes. Thermal processes	New equipment re-using waste heat	5–10% reduction, capital investment required
		Auxiliary energy (AE)	Operations idle equipment air/water human factors	Operational and behavioural changes	20% reduction low/no cost initiatives
	Indirect energy (IE)		HVAC lighting grid	New facility design/retrofit. Demand side management (DSM)	5–10% reduction, capital investment required

measures can be introduced. This allows decisions makers a more holistic view of energy usage in an industrial facility, with more focus on the potential for reduction. For mature factories where the low-hanging fruit of Tariff Structures, Lighting and VSDs have been addressed the most obvious 'sweet spot' for cost reduction is to address auxiliary energy consumption through change in operations and behaviours.

Information from an industry study [35] in Ireland established that the relative percentage of direct versus indirect energy usage on a large manufacturing sits was 57% (DE) to 43% (AE) respectively. In other words 57% of the energy consumption went towards making the products and 43% of the total energy consumption went towards supporting the production environment. The direct energy usage was then analysed into either the value added energy or the auxiliary energy. In this case study, the value added energy accounts for 31% of overall energy usage and the auxiliary energy accounts for 26% of overall energy usage. The flow of energy in the SEUs and the approximate direct, (value added and auxiliary) and indirect energy usage is diagrammatically represented using a Sankey diagram in Fig. 9.2.

9.4   Value Stream Mapping

9.4.1   Lean Energy Management

The earlier review of methodologies that categorise energy usage in industry highlighted that companies still lack appropriate methods to effectively address

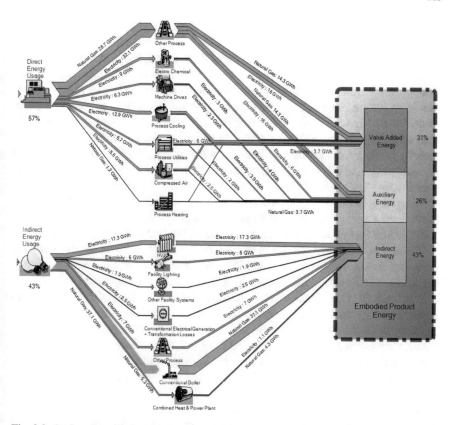

Fig. 9.2 Sankey diagram showing the flow of energy per SEU and categorised into value added, auxiliary and indirect energy

energy efficiency in a comprehensive manner. The approach proposed here introduces a practical process mapping methodology that combines energy management with value stream mapping that can clearly link the provision of technical building services (Air/Water/Steam) to the production requirements. The methodology is non-complex, applies lean manufacturing principles and upon application to a medical device discrete manufacturing facility was successful in identifying the relationship between the energy usage and production activities for a particular value stream.

Energy Management focuses on the systematic use of management and technology to improve energy performance in a selected site. It requires that energy procurement, energy efficiency and renewable energy be integrated, proactive and incorporated in order for it to be fully effective [54]. Value Stream Mapping (VSM), a widely used tool of Lean Manufacturing, is a type of symbolic model that graphically enables the end user to observe the material and information flow as product or service travels through a value chain [55]. The model represents the flow of resource such as materials, information and personnel along with their interactions through a production chain. It specifies activities and cycle times and also identifies value-added and non-value added activities in the process. It allows the visualisation of all the manufacturing system, rather than just the equipment.

Wormak and Jones [56] suggest that five principles of the Lean thinking philosophy are required for value stream mapping. Firstly, value must be defined; providing the customer with the right product or service, at the right time and price, as determined in each case by the customer. Secondly, the value stream must be determined; specifying particular activities needed to design order and distribute a specific product from concept to launch. Thirdly, tasks must be designed so that through progressive achievement, stoppages, scrap and backflow are eliminated. Fourthly, the "pull system"; a system designed to where nothing is produced by the upstream supplier until the downstream customer requires it, must be implemented. Finally, the target of complete elimination of waste should be constantly reviewed with the aim that all activities across a value stream must only create value.

The use of VSM and energy management is present in literature and has been trialed in certain industries in the US [55]. An example of this is the work carried out by the US Environment Protection Agency in the development of the "Lean, Energy and Climate Toolkit" [57]. This provides strategies and techniques to improve energy and environmental performance in tandem with achieving leans goals such as quality, reduced waste and improved customer responsiveness. Despite the fact that it provides detailed information in relation to lean principles, energy monitoring and targeting, and green-house emissions management, the output tool is still quite complicated and prior knowledge of VSM is required to understand and use it.

Based on the principle that VSMs serve as a magnifying glass to view the whole manufacturing system, Fraizer [58] has proposed the use of the "concept of value" and the VSM tool as a means of determining energy consumption in a current state. In particular, the work focuses on determining energy characteristics of the process.

Paju et al. suggest the concept of Sustainable Manufacturing Mapping (SMM). This is based on the combination of VSM, Life Cycle Assessment (LCA) and Discrete Event Simulation (DES) to provide a highly visual model that allows for the assessment of sustainability indicators in manufacturing. The main outcomes are goal definition, identification for sustainability indicators, and modelling of current and future state process maps. Despite the robustness of SMM work, the main challenges observed are the idea that a goal-oriented approach can be quite complicated, as the assessment does not use the same indicators every time to carry out an evaluation. In larger multinational companies, where each VS or SBU operate as "small factories" it could be difficult to compare performance against one another, or even set targets for the company as a whole if the indicators are not shared across the board.

Due to the complexity and prior knowledge of particular techniques for the application and implementation of the above methodologies, the process mapping methodology proposed in this chapter follows the basic principles of Value Stream

Mapping and encompasses the concept that production is multidimensional and that system dynamics are critical to the evaluation of a production area. It also includes both direct and indirect factors that affect energy efficiency in production operations.

9.4.2 Mapping Methodology

The primary aim of the proposed process mapping methodology is to effectively acquire production and energy data from a production environment that could be modelled to provide both steady-state and dynamic energy consumption and potentially provide a multi-dimensional hierarchical view of this energy consumption and cost directly related to production equipment.

The proposed methodology to acquire such data was designed around the following principles:

- The methodology is not related or restricted to a specific case but generic in nature and applicable to diverse manufacturing types (i.e. continuous and discrete).
- The methodology pursues a holistic perspective of the relationship between manufacturing processes and energy consumption, including all relevant process and energy flows as well as their interdependencies.
- The methodology is flexible so that it can be applicable to small and medium sized enterprises typically facing obstacles towards energy efficiency measures and usage of simulation.
- The methodology provides multi-dimensional evaluation of improvement measures in all relevant fields of actions.
- The methodology can adapt to an ever changing production environment such as equipment relocation or process improvement.

The methodology consists of five main steps;

1. Process Step Identification,
2. Equipment Identification,
3. Determination of Significant Energy Users,
4. Technical Services Identification and
5. Data Collection Availability.

These steps are generally applied to one value stream or strategic business unit to create a process map but can be scaled up and/or aggregated to factory level, providing the overall production process and energy usage of a factory.

1. Process Step Identification

Each process step in the production chain is identified and labelled according to production specifications or internal factory documents. Both the throughput (i.e.

batch size) and the cycle time for each process step for each unit of manufacturing (i.e. cycle time/batch) is identified. Differentiation between automated and manual steps is highlighted, as manual steps are not considered unless determined to have a significant impact.

2. Process Equipment Identification

The equipment used for each process is then identified along with the quantity of equipment per step. This is critical as there may be a one-to-many relationship between the process step and process equipment although generally each product will only take one path through the process. Process equipment energy consumption data is then collected. The electrical consumption provided by the manufacturer (typically referenced on the equipment plate or manuals) should be collected, as well as any thermal energy usage (i.e. gas to generate process heat).

3. Determination of Significant Energy Users

Based on the cycle time and the energy consumption data of each item of equipment, a list of process SEUs can be determined. It is critical to take into account the accurate cycle times, as the machine rating alone may not be suitable to assess the scale of the energy consumption involved from a product perspective.

4. Technical Services Identification

It is necessary to identify the technical services (compressed air, water, steam, nitrogen, dust extraction, etc.) used by each process step. These services require both electrical and thermal energy and should be accounted for as part as the energy usage of the process. As specific metering at the process step is not usually available, a method of allocating consumption of the technical services across the value stream must be developed.

5. Data Collection Availability

It is necessary to identify if there is sub-metering available at process level for both electrical and thermal energy. If energy meters are installed at this step then information can be gathered from the energy monitoring system. If meters are not in place, then it may be possible to use control information from variable speed drives (VSDs) or programmable logic controllers (PLCs) on the machines or to deploy sensors that can gather data on the behaviour of the process (cycle time, temperature, etc.).

By applying the above five steps to a production environment, the relationship between production and the energy consumption in the technical services and process steps is highlighted. This can be used to understand how manufacturing activities function within an industrial facility and how energy and manufacturing are interrelated. It also identifies the true significant energy users for a particular process and can highlight both the value added and auxiliary energy, where energy optimisation and reduction techniques can be applied. A sample high level process map carried out on a value stream is outlined in Fig. 9.3.

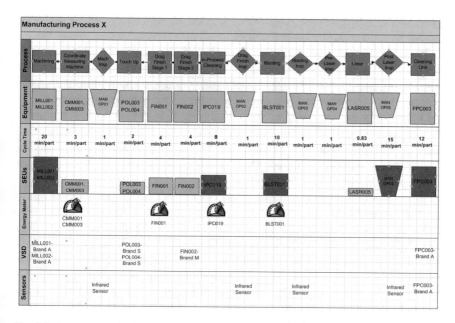

Fig. 9.3 Process mapping methodology output visual representation

As is clear in row 4, the items in red represent Process SEUs and their respective size reflects the scale of the energy consumption. Row 5 show the existing meters available and it is clear that not all the SEUs are appropriately monitored. Additional rows can then be added to reflect the different cross-cutting or technical services that apply at each significant production step.

The method is one that can be updated easily to reflect changes in the production environment and to provide a holistic view of the energy and technical services in the context of the varying production activity.

9.5   Conclusion

The chapter describes the background to energy usage in production operations and sets out some principles, process steps and methods to provide a more holistic view of the Significant Energy Users (SEUs) and the related consumption of energy and technical services (Air, Water, etc.).

Sustainable Manufacturing (SM) is the new paradigm for manufacturing companies and involves the integration of all relevant dimensions that affect or have effects on third parties while conducting manufacturing operations, including energy, environmental impact and life-cycle analysis. Roadmapping exercises with European Industries have described the smart sustainable factory of the future will be one where there is full integration between the production activity and the associated energy used and where the operation of the factory can be optimised around its energy and ecological impact.

Potential reductions in energy consumption of 20% and in GHG emissions of 30% have been identified on industrial sites where there is an in-depth understanding of energy flows in the manufacturing process and a clear analysis of energy usage. However multiple studies have found that actual savings of 10–12% are more likely due to identified barriers such as; lack of time, lack of resources, lack of knowledge and a primary focus on production. Some changes in industrial practice is required to achieve the full potential savings that are available, including the analysis of NonEnergy Benefit (NEBs) which can be worth up to 2.5 times the energy efficiency improvements alone.

A model based on direct and indirect energy analysis from a 'product' viewpoint has been extended to identify waste or auxiliary energy in line with 'Lean' principles in manufacturing. The methodology outlines the process flow, energy metering requirements, the technical utilities servicing the process and an identification of the significant energy users. In large industrial facilities it has been shown that up to 60% of energy consumption is directly consumed in production activities, although of this, it has been shown that anywhere from 52–85% of the energy may be used for functions that are not directly related to the production of parts. In one application on an industrial site in Ireland, auxiliary energy of 26% was identified. This auxiliary energy identified represents the best opportunity to gain energy savings through operational and behavioral changes at the lowest possible cost.

The key energy consumers (SEUs) that need close attention on industrial sites normally includes the following;

- Heating Systems for facility and provision of hot water/steam for production processes
- Compressed Air systems
- CHP and waste heat recovery
- Provision of chilled water and di-ionised water
- Facility internal and external lighting
- Electrical tariffs, structures and conversion/distribution
- Production machines and conveyors
- Operating Procedures/Employee Behaviour
- Air Conditioning/Vacuum Extraction

In addition, due to the significance (over 70%) of electricity consumption in motive power and potential savings of 20–30%, industry should have a clear strategy on the use of energy efficiency motors, on the correct sizing of motors, on the use of variable speed drives (VSDs) and on the use of advanced control systems to optimise performance.

Developing a clear link between the temporal profile and/or efficiency of the energy consumption by the specific value stream can provide full transparency in the impacts (costs, emissions) of energy consumption and can provide positive feedback and cost reduction to reward improved performance by the value stream. The ability to link Production and Energy models is also a vital link in the future application of demand side management to industry.

The proposed process mapping methodology [Value Stream Mapping (VSM)] effectively acquires production and energy data that could be modelled to provide both steady-state and dynamic energy consumption and potentially provide a multidimensional hierarchical view of this energy consumption and cost directly related to production equipment. The approach of using a 'product' viewpoint is very much in line with the now standard approach of 'Lean' manufacturing and the method is one that can be updated easily to reflect changes in the production environment and to provide a holistic view of the energy and technical services in the context of the varying production activity.

Acknowledgements The input and assistance of researchers, Thomas O'Connor and Maria Rivas, is gratefully acknowledged and appreciated.

The research work is supported by Enterprise Ireland (EI), the Sustainable Energy Authority of Ireland (SEAI), Science Foundation Ireland (SFI), and the Industrial Development Agency (IDA Ireland) and has been carried out in collaboration with Limerick Institute of Technology (LIT), University of Ulster (UU), Innovation for Irelands Energy Efficiency Research Centre (i2e2) and the International Energy Research Centre (IERC).

References

1. IEA: World Energy Outlook (2013)
2. Eurostat Statistical Database, 2014. http://ec.europa.eu/eurostat/data/database. Accessed 13/03/2015
3. Jens, K.: Energieeffiziente Produktherstellung. Fraunhofer FUTUR Produktionstechnik für die Zukunft. **12**, 12–13 (2010)
4. Digest of United Kingdom Energy Statistics (DUKES), 2013 (Internet Booklet). Department of Energy and Climate Change, TSO. https://www.gov.uk/government/uploads/attachment_data/file/225056/DUKES_2013_internet_booklet.pdf. Accessed, May 2014
5. Herrmann, C., Thiede, S., Heinemann, T.: A holistic framework for increasing energy and resource efficiency in manufacturing. In: 8th Global Conference on Sustainable Manufacturing, Abu Dhabi, pp. 267–273 (2011)
6. European Commission: Energy efficiency in manufacturing—the role of ICT. Consultation Group on Smart Manufacturing. Publication of the European Communities. ISBN 978-92-79-11306-2, Feb 2009
7. Paju, M., Heilala, J., Hentula, M., Heikkila, A., Johansson, B., Leong, S., et al.: Framework and indicators for a sustainable manufacturing mapping methodology. In: 2010 Winter Simulation Conference, Baltimore, USA, pp. 3411–3422 (2010)
8. Energy Savings Toolbox—an Energy Audit Manual and Tool, Natural Resources Canada. http://www.nrcan.gc.ca/energy/efficiency/industry/cipec/5161, Accessed 10th Jan 2015
9. Rahimifard, S., Seow, Y., Childs, T.: Minimising embodied product energy to support energy efficient manufacturing. CIRP Ann Manuf Technol **59**, 25–28 (2010)
10. FoF: Factories of the future PPP, strategic multi-annual roadmap. Publications Office of the European Union, Luxembourg (2010). ISBN 978-92-79-15227-6

11. The U.S. Department of Commerce's Sustainable Manufacturing Initiative and Public-Private Dialogue. Available: http://trade.gov/competitiveness/sustainablemanufacturing/ June, 2013. Accessed 20/01/2015
12. OECD Sustainable Manufacturing Toolkit—Seven Steps to Environmental Excellence. http://www.oecd.org/innovation/green/toolkit/48661768.pdf. Accessed 20th May 2014
13. Costello, G.J., Lohan, J., Donnellan, B.: Mobilising IS to support the diffusion of energy management practices outside of Ireland's LIEN (Large Industry Energy Network). In: Proceeding of the 15th Annual Conference of the Association Information et Management (AIM 2010), IDDS, La Rochelle, France, May 2010
14. The UK Carbon Trust—Industrial Energy Efficiency Accelerator (IEEA) Programme, 2014. http://www.carbontrust.com/client-services/technology/innovation/industrial-energy-efficiency-accelerator. Accessed May 2014
15. Tracking industrial energy efficiency and CO_2 emission. IEA, June 2007. ISBN: 978-92-64-03016-9
16. Besparelser i erhvervslivet; Viegand & Maagøe for the Danish Energy Agency; Feb 2010
17. IEA and IIP 2012; IEA (International Energy Agency) and IIP (Institute for Industrial Productivity): Energy management programmes for industry: gaining through saving. IEA Publications, Paris, France (2012). https://www.iea.org/publications/freepublications/publication/policypathwaysindus¬try.pdf. Accessed 25th June 2014
18. Duarte, C., Acker, B., Grosshans, R., Manic, M., Van Den Wymelenberg, K., Rieger, C.: Prioritizing and visualizing energy management and control system data to provide actionable information for building operators. In: Western Energy Policy Research Conference, Boise, ID, U.S. 25–26 Aug 2011
19. National Academy of Sciences (NAS): Real Prospects for Energy Efficiency in the United States. National Academy of Sciences, Washington, DC (2010)
20. McKinsey & Company: Unlocking energy efficiency in the US economy, June 2009. http://www.mckinsey.com/client_service/electric_power_and_natural_gas/latest_thinking/unlocking_energy_efficiency_in_the_us_economy. Accessed 13/03/2015
21. Regional Greenhouse Gas Initiative: Regional investment of RGGI CO_2 allowance proceeds, 2012. http://www.rggi.org/design/program-review. Accessed 13/03/2015
22. Non-energy benefits from commercial and industrial energy efficiency programs: Energy efficiency may not be the best story. Paper Presented at the Energy program Evaluation Conference, Seattle, Nick P. Hall & Johna A. Roth, Tec-Market Works (2003)
23. Jollands, N., Tanaka K., Gasc, E., Wescott, W., Energy management action network (EMAK)—a scoping study investigating the establishment and support of an international and domestic action network of energy management in industry. International Energy Agency, France (2009)
24. Granade, C., Creyts, J., Derkach, A., Farase, P., Nyquist, S., Ostrowski, K.: Unlocking energy efficiency in the U.S. economy. McKinsey & Company, USA (2009)
25. Christoffersen, L.B., Larsen, A., Togeby, M.: Empirical analysis of energy management in Danish industry. J. Clean. Prod. 14(5), 516–526 (2006)
26. Thollander, P., Ottosson, M.: Energy management practices in Swedish energy-intensive industries. J. Clean. Prod. 18(12), 1125–1133 (2010)
27. Thollander, P., Palm, J.: Improving energy efficiency in industrial energy systems—an interdisciplinary perspective on barriers, energy audits, energy management, policies and programs. Springer, Berlin. ISBN 978-1- 4471-4161-7 (2012)
28. Seow, Y., Rahimifard, S.: A framework for modelling energy consumption within manufacturing systems. CIRP J. Manuf. Sci. Technol. 4, 258–264 (2011)
29. Cornelis, E.: Lessons learnt from two long-term agreements on energy-efficiency in industry in Flanders, Belgium. In: Proceedings of the ECEEE 2014 Summer Study on Energy Efficiency in Industry. Arnhem, 3–5 June 2014
30. Backlund, S., Paramonova, S., Thollander, P., Rohdin, P., Karlsson, M.: A regional method for increased resource-efficiency in industrial energy systems. In: Proceedings of the ECEEE 2014 Summer Study on Energy Efficiency in Industry. Arnhem, 3–5 June 2014

31. Perkins, J., Maxwell, J.: Energy audit impacts delivering sustained savings. In: Proceedings of the ECEEE 2014 Summer Study on Energy Efficiency in Industry. Arnhem, 3–5 June 2014
32. Müller, E., Löffler, T.: Improving energy efficiency in manufacturing plants case studies and guidelines. In: 16th CIRP International Conference on Life Cycle Engineering (LCE 2009), pp. 465–471. Department of Factory Planning and Factory Management, Chemnitz University of Technology, Saxony, Germany (2009)
33. Seow, Y., Rahimifard, S.: A framework for modelling energy consumption within manufacturing systems. CIRP J. Manufact. Sci. Technol. **4**, 258–264 (2011)
34. Sorrell, S., Mallett, A,. Nye, S.: Barriers to industrial energy efficiency: a literature review. United National Industrial Development Organisation (UNIDO), Development Policy, Statistics and Research Branch, Vienna, 10/2011
35. Harrington, J., Cosgrove, J., Ryan P.: A strategic review of energy management systems in significant industrial sites in Ireland. In: Conference on Energy Efficiency in Industry, European Council for an Energy Efficient Economy (ECEEE), Arnheim, NL. 3–5 June 2014
36. Epstein, G., D'Antonio, M., Neiman, L., Perkins J.: Large industrials: serious engagement for deep savings. In: Proceedings of the ECEEE 2014 Summer Study on Energy Efficiency in Industry, Arnhem, 3–5 June 2014
37. I2E2—Industrial Innovation for Energy Efficiency. http://www.i2e2.ie/. Accessed 15/03/2015
38. SESEC, Sustainable Energy Savings for the European Clothing Industry (SESEC). http://euratex.eu/pages/sesec. Accessed 20th Jan 2015
39. SurfEnergy-Path to Energy Efficiency. https://ec.europa.eu/energy/intelligent/projects/en/projects/surfenergy. Accessed 20th Jan 2015
40. Deshpande, A., Kumar, S., Tulsyan, A.: An analysis of the best practices in energy efficiency adopted by the Indian manufacturing sector. In: Conference on Energy Efficiency in Industry, European Council for an Energy Efficient Economy (ECEEE), Arnheim, NL. Sept 2012
41. University of Daytona Industrial Assessment Centre (UDIAC). https://www.udayton.edu/engineering/centers/industrial_assessment/index.php. Accessed 20th June 2013
42. Energy Savings Toolbox—an Energy Audit Manual and Tool. Natural Resources Canada, http://www.nrcan.gc.ca/energy/efficiency/industry/cipec/5161. Accessed 10th January 2015
43. Brunner, C., Waide, R.: Energy efficiency policy opportunities for electric motor-driven systems. Working Pager, IEA, OECD, Paris (2011)
44. US Department of Energy: Annual Energy Outlook 2011. US Energy Information Administration, US Department of Energy
45. Rita Werle, R., Conrad, U., Brunner, C., Cooremans, C.: Financial incentive program for efficient motors in Switzerland: lessons learned. In: Proceedings of the ECEEE 2014 Summer Study on Energy Efficiency in Industry. Arnhem, 3–5 June 2014
46. Schmid: Energieeffizienz in Unternehmen - Eine wissensbasierte Analyse von Einflussfaktoren und Instrumenten. ETH Zürich: vdf Hochschulverlag AG (2008)
47. Giacone, E., Mancò, S.: Energy efficiency measurement in industrial processes. Energy **38**, 331–345 (2012)
48. NAM: Efficiency and innovation in U.S. manufacturing energy use. National Association of Manufacturers, NAM (2006)
49. Dahmus, J.B., Gutowsky, T.C.: An environmental analysis of machining. In: ASME International Mechanical Engineering Congress and RD&D Expo, Anaheim, California, USA (2004)
50. Drake, R., Yildirim, M.B., Twomey, J., Whitman, L., Ahmad, J., Lodhia, P.: Data collection framework on energy consumption in manufacturing. In: IEEE Annual Conference and Expo, Orlando, Florida, USA (2006)
51. Seow, Y., Rahimifard, S.: Improving product design based on energy considerations, globalized solutions for sustainability in manufacturing. In: 18th CIRP International Conference on Life Cycle Engineering, Braunschweig, Germany, pp. 154–159 (2011)
52. Mustafaraj, G., Cosgrove, J., Rivas, M.: A methodology for determining auxiliary and value-added electricity in manufacturing machines. Int. J. Prod. Res. (International Foundation for Production Research, Taylor & Francis, Sept 2015, in Press)

53. Bunse, K., Vodicka, M., Schönsleben, P., Brülhart, M., Ernst, F.O.: Integrating energy efficiency performance in production management—gap analysis between industrial needs and scientific literature. J. Clean. Prod. **19**, 667–679, 2011/5/2011
54. Carbon Trust. Energy management—a comprehensive guide to controlling energy use (2011). http://www.gbc.ee/710eng.pdf. Accessed 15/03/2015
55. Keskin, C., Kayakutlu, G.: Value stream maps for industrial energy efficiency. In: PICMET'12: Technology Management of Emerging Technologies, pp. 2834–2831. Vancouver, Canada (2012)
56. Wormack, J.P., Jones, D.T.: Lean Thinking. Simon & Schuster, New York (1996)
57. EPA: Lean, Energy, and Climate Toolkit. U. S. E. P. Agency (2011)
58. Fraizer, R.C.: Bandwidth analysis, lean methods, and decision science to select energy management projects in manufacturing. Energy Eng. **105** (2008)

Chapter 10
Two-Stage Optimization for Building Energy Management

Jorn K. Gruber and Milan Prodanovic

Abstract In recent years, the energy sector has undergone an important transformation as a result of technological progress and socioeconomic development. The continuous integration of renewable technologies drives the gradual transition from the traditional business model based on a reduced number of large power plants to a more decentralized energy production. The increasing energy demand and intermittent generation of renewable energy sources require modern control strategies to provide an uninterrupted service and guarantee high energy efficiency. Utilities and network operators permanently supervise production facilities and grids to compensate any mismatch between production and consumption. The enormous potential of local energy management contributes to grid stability and can be used to reduce the adverse effects of load variations and production fluctuations. This paper presents a building energy management which determines the optimal scheduling of all components of the local energy system. The two-stage optimization is based on a receding horizon strategy and minimizes two economic functions subject to the physical system constraints. The performance of the proposed building energy management is validated in simulations and the results are compared to the ones obtained with other energy management approaches.

Keywords Building energy management · Demand response · Two-stage optimization

J.K. Gruber (✉) · M. Prodanovic
Electrical Systems Unit, IMDEA Energy Institute, 28935 Móstoles, Spain
e-mail: jorn.gruber@imdea.org

M. Prodanovic
e-mail: milan.prodanovic@imdea.org

10.1 Introduction

Continuity and security of energy supply are the two most important factors behind industrial development and are closely related to the prosperity of a society. The continuity is achieved though permanent planning, supervision and control of generation facilities and transmission systems. Utilities and network operators have to compensate any mismatch between generation and consumption in order to maintain power quality and avoid service failures. In recent years grid balancing has become a challenging issue because of the massive integration of intermittent renewable energy sources exposing the limitations of the infrastructure used for energy production, transmission and distribution. The gradual shift from a centralized structure towards a more decentralized and distributed system with a partially unpredictable production requires an advanced energy management to maintain grid stability and ensure an uninterrupted energy supply. Deployment of smart meters and availability of affordable and reliable energy storage technologies convert buildings and their energy management systems in an interesting tool for providing demand flexibility. Buildings contribute significantly to the total energy demand—in some countries up to 45% of the primary energy consumption [1–3]—and can be used to actively participate in the grid balancing issues.

Building energy management is considered a feasible approach to reduce the effects of intermittent generation and demand variations [4, 5]. Demand response and load shaping techniques provide customers with the facility to contribute to system balancing. Advanced energy management on the consumer side enables a closer tracking of the energy production, i.e. the traditional load following approach is inverted, and can be used to minimize the mismatch between generation and consumption.

Nowadays, energy is usually offered to customers at fixed rates without considering actual costs for generation, transmission and distribution. Real-time pricing and time-of-use (TOU) tariffs provide utilities, network operators and retailers with an option to pass on the real energy costs to the customer [6–8]. A direct link between wholesale and retail market is a necessary condition for a dynamic, efficient and competitive energy market. Variable pricing schemes in building energy management can be used to incentivize customers to shift their energy consumption from peak times and to periods of low energy demand.

The growing interest both from the industrial sector and the scientific community led to an increasing research in building energy management. A large number of the developed techniques rely on the minimization of resulting energy costs, including receding horizon strategies based on model predictive control (MPC) [9, 10], multi-agent systems [11, 12], mixed integer linear programming [13, 14] and game theory [15]. In contrast, the maximization of the reward obtained from a temporal reduction of the amount of energy drawn from the electric grid has been proposed in a demand bidding program for hotel energy management [16]. The participants may respond to the external bid and reduce their energy consumption or operate as usual without any monetary savings or other benefits. The home energy management proposed in [17] contains a multi-objective approach for stability enhancement, long-term performance optimization and energy trading. The strategy developed in [18] uses a genetic algorithm to minimize both energy costs and appliance operation delays. Besides, this approach enables the reduction of the power peak-to-average and leads to an improved grid stability.

Two- and multi-stage techniques are frequently employed in energy management to reduce the mathematical complexity of the optimization problem. Many of these strategies use predictions to determine the optimal time-ahead scheduling and a real-time closed loop control to compensate any deviation resulting from uncertainties [19, 20]. The building energy management in [21] considers the random local generation as an uncertainty with unreliable probabilistic information. The proposed robust optimization solves the resulting constrained worst-case scenario using an two-stage approach based on mixed integer programming. The demand response strategies presented in [22, 23] maximize utility benefit and customer welfare in separate stages. The optimal energy price in the proposed multi-objective approach is determined by means of game-theoretic and iterative algorithms which ensure convergence.

This work presents a building energy management based on a two-stage optimization. The approach divides the optimization problem in two less complex problems with different prediction horizons and sampling times. The energy management considers the physical constraints of the system and computes the optimal scheduling by minimizing an economic index in each optimization stage. The optimization problems, formulated as mixed integer linear programs (MILP), can be solved by suitable solvers.

The paper is organized as follows: Sect. 10.2 gives a general description of building energy management and describes the general architecture of the proposed strategy. Section 10.3 presents the two-stage optimization approach and explains the computation of the optimal scheduling. The definition and implementation of a case study, a medium size hotel with photovoltaic panels and battery system, is given in Sect. 10.4. Section 10.5 presents some simulation results and compares the performance of the proposed approach to the one of other energy management strategies. Finally, the most important conclusions are drawn in Sect. 10.6.

10.2 Building Energy Management

In recent years, energy management on the demand side has captured notable attention of the scientific community and many companies from the energy sector. The application of building energy management can benefit customers, network operators and utilities.

10.2.1 Problem Description

Modern energy management has to ensure a continuous supply, energy quality, grid stability and security of the used infrastructure. Traditionally, utilities and transmission network operators used the load-following approach to compensate any mismatch between generation and consumption. The massive integration of renewable energy sources increased the complexity of the energy system and led to a more distributed and decentralized generation. These changes motivated an intensified research on local energy management both in companies and scientific communities. The enormous potential of local energy management, including building energy management, allows reducing adverse effects of load variations and intermittent generation. Modern techniques such as demand response and load shaping enable close tracking of generation and can improve grid stability and energy quality.

A building energy management determines the scheduling for all components of the local energy system. The computation of the optimal solution combines local and global aspects of the building energy system. Common elements considered by the optimization procedure include the customer demand, local generation, grid connection and storage capacities (see Fig. 10.1). Dynamic pricing schemes or time-of-use (TOU) tariffs are frequently employed to motivate customers to change their consumption habits.

The local energy management usually computes the scheduling by solving an optimization problem. These optimization problems consist of an objective function and consider different constraints such as the energy system's physical limitations. If the objective function is based on the energy supply costs, the optimal solution represents the economically most reasonable scheduling.

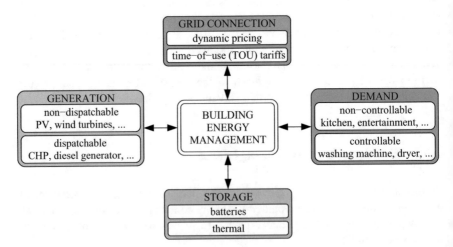

Fig. 10.1 General scheme of the key elements considered in a demand side management approach

An important issue in the development of a building energy management is the choice of the optimization horizon and sampling time. The use of long scheduling horizons and low sampling times leads to highly complex optimization problems. In the case of energy systems with many resources the computational burden prevents the use of local energy management on systems with reduced computation power such as smart meters or low-cost embedded systems. The development of advanced strategies which can be used on such devices will facilitate the success of building energy management. The choice of the optimization horizon and the sampling time represents a trade-off that ensures fast optimization and good scheduling.

10.2.2   Two-Stage Optimization Framework

The proposed building energy management uses an advanced optimization approach to compute the optimal scheduling for the energy system. The optimization considers all components affecting the energy balance, including dispatchable generation systems (e.g. diesel generators), non-dispatchable generation (e.g. photovoltaic installations), grid connection, storage devices, non-controllable loads and controllable loads. The controllable loads can be divided into two categories: deferrable loads which can be shifted freely in a given period and power-level controllable loads which permit certain variation in the power around the nominal value (e.g. between 90 and 110%).

The developed approach divides the optimization problem into two less complex problems: a medium term optimization problem over one day and a short term optimization problem with a horizon of one hour (see Fig. 10.2). The two-stage optimization solves in a first step the medium term problem and employs the obtained solution to define the short term problem. In a second step, the optimization approach solves the short term problem and applies the resulting optimal scheduling to the building energy system.

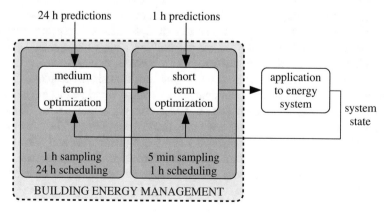

Fig. 10.2 General structure of the proposed two-stage optimization approach

The optimization problems are defined using the system state—the state of charge (SOC) of the storage systems—and the predictions of the demand and generation. The system state introduces some kind of memory to the energy system and interconnects both optimization problems. The proposed energy management has been developed for devices with a reduced computation power and can be used in an integrated building management system (BMS).

10.3 Optimization Approach

The proposed building energy management computes the scheduling of all components of the building energy system by means of a two-stage optimization. The following sections describe the general idea of the energy management as well as the two stages of the optimization approach.

10.3.1 Overview

The building energy management combines the available information from internal and external sources and computes the solution to the underlying problem. The obtained solution contains the optimal scheduling for all elements of the considered energy system and minimizes the corresponding energy costs. The used two-stage optimization (see Fig. 10.2) has been developed to ensure fast optimization and good scheduling.

In the first stage a medium term optimization problem defined over a horizon of one day and with the sampling time of one hour is solved. In the second stage, the results obtained in the medium term optimization are used to define the short term optimization problem. The solution to this problem contains the optimal scheduling over a horizon of one hour with a sampling time of five minutes. From the obtained short term solution only the scheduling for the current sample is applied to the energy system. In the next sample, the short term optimization is repeated by using updated values. After 12 samples, i.e. 1 h, the building energy management solves a new medium term optimization problem over a horizon of 1 day.

The energy management has been developed for its use in the residential sector on hardware with a limited computational power such as smart meters or embedded systems. The division in two optimization problems in combination with the chosen sampling times leads to a relatively low mathematical complexity. The use of lower sampling times would require improved time-ahead predictions and increase the complexity of the resulting optimization problems. Possible deviations in the demand or generation with respect to the used predictions can be compensated using real-time control strategies.

10.3.2 Medium Term Optimization

The medium term optimization computes the optimal scheduling for the considered energy system over a horizon of one day ($h^m = 1$ d) with a sampling time of one hour ($t^m = 1$ h). The one-day-ahead predictions of the different loads and non-dispatchable generation, the state of charge (SOC) of the storage devices as well as the energy prices of the different sources (including negative prices for possible benefits in the case of an energy export to the main grid) are employed to define the optimization problem, see Fig. 10.3.

The solution to the problem contains the optimal scheduling for the deferrable loads, storage devices, dispatchable generation and grid connection. The scheduling of the storage and the deferrable loads is employed in the definition of the short term optimization problem and is not recalculated in the second stage of the optimization. This method avoid frequent changes in the power of the storage devices and the deferrable loads and guarantees a constant value during an entire hour. In contrast, the obtained medium-term scheduling for the dispatchable generation and the grid connection is discarded and recalculated in the short term optimization.

In the medium term optimization the power-level controllable loads are considered as non-controllable loads. Using the nominal values, the power-level controllable loads are not optimized at this stage of the building energy management. The optimization of these loads is carried out later in the short term optimization (see Sect. 10.3.3).

10.3.3 Short Term Optimization

The second stage of the proposed building energy management determines the optimal scheduling over a horizon of one hour ($h^s = 1$ h) with a sampling of five minutes ($t^s = 5$ min), see Fig. 10.4. The short term optimization problem is defined using the following information:

Fig. 10.3 Scheme of the medium term optimization with a horizon of one day and a sampling time of 1 h

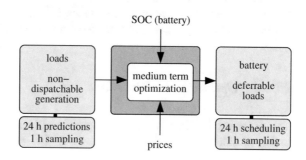

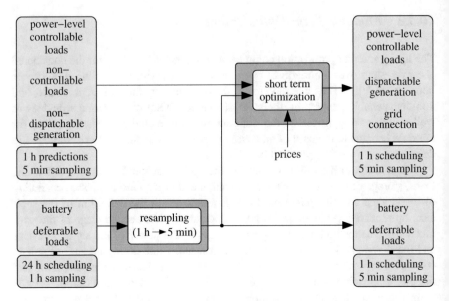

Fig. 10.4 Scheme of the short term optimization with a horizon of hour and a sampling time of 5 min

- medium term scheduling: storage devices and deferrable loads (see Sect. 10.3.2)
- one-hour-ahead predictions: power-level controllable loads, non-controllable loads and non-dispatchable generation
- energy prices

The solution to the short term optimization problem consists of the optimal scheduling for the dispatchable generation, grid connection and power-level controllable loads. The short term scheduling is recalculated every 5 min, i.e. the short term optimization is executed 12 times between two successive medium term optimizations.

Note that the short term optimization does not determine a new scheduling for the battery and the deferrable loads. For these components the results obtained in the medium term optimization are resampled for the sampling time of 5 min, i.e. the power is kept constant between two medium term optimizations.

10.3.4 Optimization Procedure

The proposed building energy management solves two optimization problems under consideration of different constraints. The minimization of the objective function returns the optimal scheduling of the building energy system. The optimal solution u^* of the proposed building energy management can be written in general form as:

$$u^* = \arg \min_{u} F(u)$$

$$\text{s.t. } g(u) \leq 0 \tag{10.1}$$

$$h(u) = 0$$

where u consists of discrete and continuous decision variables and $F(u)$ is the cost function. The functions $g(u)$ and $h(u)$ are used to define inequality and equality constraints, respectively. In order to obtain optimization problems with a low mathematical complexity, only linear objective functions and constraints have been employed. The optimization problems have the structure of mixed integer linear programs (MILP) [24, 25] and can be solved by appropriate numerical solvers.

The objective functions used in the building energy management are based on the energy supply costs. The optimal solution represents the economically most reasonable scheduling for the energy system. From the obtained solution, only the values for the current sample are applied to the system. The employed receding horizon strategy (frequently employed in model predictive control [26–28]) computes in the next sample a new solution with updated predictions.

The accuracy of the predictions used in the two optimization stages has an impact on the quality of the scheduling of the energy system. The frequent recalculation of the optimal solution provides an improved energy management in the presence of uncertain predictions. The use of a power reserve in combination with soft constraints helps to stabilize the energy system and to compensate the effect of large deviations between predictions and actual values.

10.4 Case Study

The performance of the developed building energy management was verified in a simulated case study. The characteristics of the considered energy system and the implementation of the energy management will be described in this following paragraphs.

10.4.1 Building Energy System

A small to medium size hotel has been chosen to test the building energy management. The energy system is used to validate the proposed two-stage optimization approach.

For the energy supply, the hotel has been equipped with two diesel generators, a bidirectional grid connection, photovoltaic (PV) panels and a battery system (see Fig. 10.5). The installed diesel generators have a nominal power of 30 kW and

10 kW, the grid allows an energy exchange between −30 kW (export) and 30 kW (import) and the PV modules have a nominal power of 12 kW. The battery has a capacity of 63 kWh, but in order to avoid an excessive degradation the storage is charged/discharged only in the interval from 20% (12.6 kWh) to 80% (50.4 kWh). Furthermore, the permitted maximum power of the battery is limited to −7.6 kW (charging) and 7.6 kW (discharging).

Depending on the energy source, different energy prices have been considered. In the case of the two diesel generators, the price of 0.5 €/kWh is used. The energy production by the photovoltaic panels does not generate additional costs, i.e. the price is 0 €/kWh. The degradation of the battery has been considered using prices of 0.005 €/kWh both for charging and discharging. Efficiencies of 0.95 have been used for the process of battery charging and discharging, respectively. In the case of the grid connection, the energy excess is refunded with −0.04 €/kWh and for the energy import a super off-peak tariff (see Fig. 10.6 for the temporal profile) with the prices of 0.204 €/kWh, 0.152 €/kWh and 0.057 €/kWh is considered.

On the demand side, some of the most common loads of a hotel have been considered in the case study. The different loads can be divided in the following groups: non-controllable (kitchen, hotel rooms, and others), power-level control-lable (lights and air condition/ventilation) and deferrable (laundry). The power level controllable loads can be varied between 90 and 110% of the nominal values. An additional cost of 0.4 €/kWh for a reduction or an increase of the power level avoids undesired variations in the corresponding loads. The predefined time of operation of the laundry lies between 2 pm and 9 pm with a constant energy demand during the 7 h of operation. In the case of the proposed energy management, the laundry can be operated in the interval from 10 am to 9 pm. It is important to note that the power-level of the laundry cannot be modified, i.e. the laundry will be operated during 7 h without the possibility to stretch the load over the admissible period of 11 h.

Fig. 10.5 Energy system of the simulated hotel used to verify the developed building energy management

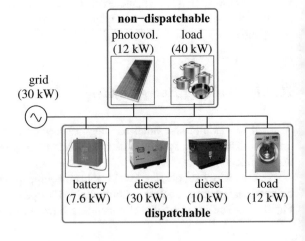

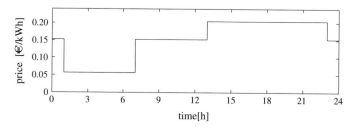

Fig. 10.6 Super off-peak tariff for the energy import from the grid considered in the case study

10.4.2 *Implementation*

The building energy management with the two-stage optimization was implemented in the Matlab environment (release 2012a) in combination with additional software packages. The advanced modelling language provided by the YALMIP toolbox [29] has been used to formulate the optimization problems. The chosen SCIP (Solving Constraint Integer Programs) solver [30, 31] offers a framework to handle a wide range of optimization problems, including the MILP problems of the proposed building energy management system. Finally, the OPTI toolbox [32] is used as an interface between Matlab and YALMIP on one hand and the SCIP solver on the other.

The proposed building energy management is implemented on a personal computer running the 32-bit version of the operating system Windows 7 with Service Pack 1. The hardware of the computer is based on an Intel Core i3-2330 M processor at 2.20 GHz with 4 GB of memory RAM.

10.5   Simulation Results

The energy management was applied in simulations to the building energy system described in Sect. 10.4.1. For the purpose of comparison the simulations were repeated with two other building energy management strategies:

- Simple energy balancing strategy: only real-time power measurements are used without considering any predictions. This energy management uses as active elements only on the grid, the battery and the diesel generators, i.e. the loads of the system are not modified by this strategy. The energy balancing uses the grid as the primary source, then the battery and finally the diesel generators. Any excess energy is firstly charged to the battery charging and then injected into the grid,
- Strategy without demand flexibility: this strategy is based on the same principles as the proposed building energy management (same predictions, horizons and sampling times), but does not include any active demand

management, i.e. only the grid, battery and dispatchable generation are available for energy management.

The first simulation of the hotel energy system was carried out with a typical energy profile with high non-controllable loads (up to 30 kW) during breakfast and dinner times. The controllable loads reach values up to 12 kW and include the deferrable load with a predefined operation between 2 pm and 9 pm and an admissible interval from 10 am to 9 pm. Besides, the photovoltaic system generates throughout the entire period of solar radiation with more than 10 kW from 8 am to 4 pm and a maximum of 12 kW at 10 am.

The results obtained with the proposed energy demand management and the simple energy balancing strategy are shown in Fig. 10.7. It can be observed that the simple energy balancing strategy requires both diesel generators (see Fig. 10.7c, d) in the evening hours to satisfy the demand (see Fig. 10.7a, b) of the energy system. In contrast, the proposed energy management strategy shifts part of the controllable load (i.e. the laundry, see Fig. 10.7a) from 2 pm to 10 am. The combination of the grid connection (see Fig. 10.7g) and the battery (see Fig. 10.7h) allows to satisfy the demand during the entire day without the use of the diesel generators (see Fig. 10.7c, d).

A comparison of the results obtained with the proposed energy management and the strategy without demand flexibility is given in Fig. 10.8. The strategy without demand flexibility uses the predictions of the loads (see Fig. 10.8a, b) and the non-dispatchable generation (see Fig. 10.8e) to compute the optimal solution. It can be observed in the evening hours that part of the demand has to be covered by the 10 kW diesel generator (see Fig. 10.8c). The proposed energy management reduces the demand in the evening by shifting part of the controllable load (see Fig. 10.8a) to an earlier hour. This modification allows satisfying the demand using the non-dispatchable generation (see Fig. 10.8e), the grid connection (see Fig. 10.8g) and the battery (see Fig. 10.8h) without using the diesel generators (see Fig. 10.8c, d).

In the simulations with the typical demand profile of a hotel and the considered energy prices for the different sources (see Sect. 10.4.1), the energy costs for the simple strategy are 65.88 €, i.e. 0.204 €/kWh. The energy costs of 55.15 € obtained with the strategy without demand flexibility led to an energy price of 0.171 €/kWh. The application of the building energy management resulted in energy costs of 46.31 € and an energy price to 0.144 €/kWh. The proposed strategy reduced the energy price by 29.4 and 15.8% with respect to the simple energy balancing approach and the energy management without demand flexibility.

The second simulation of the hotel energy system evaluates the performance of the proposed strategy with respect to changes in the predicted demand and generation. During the simulation the photovoltaic panels (non-dispatchable generation) generate only half of the predicted power. At 12 pm the energy management receives updated predictions for the non-controllable loads with a significantly higher demand (increase of 30%) from 6 pm to 11 pm, i.e. the changed values in the loads can be considered in the optimization several hours in advance.

A comparison of the results obtained with the simple energy balancing approach and the proposed building energy management is given in Fig. 10.9. It can be

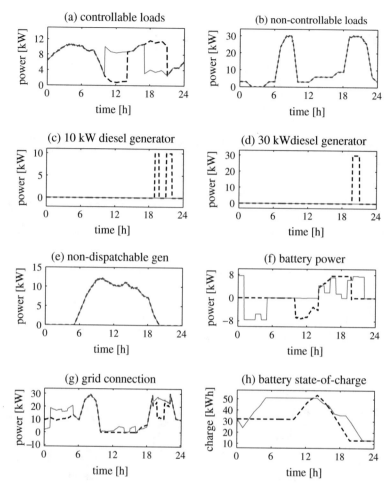

Fig. 10.7 Comparison of the simulation results obtained with the proposed building energy management (*solid line*) and the simple energy balancing strategy (*dash-dotted line*) for a typical demand profile of the considered hotel

observed that the simple energy balancing strategy hardly charges the battery (see Fig. 10.9f, g) as a result of the lower non-dispatchable generation (see Fig. 10.9e) and the reduced excess energy. Both diesel generators (see Fig. 10.9c, d) are used to cover the increased peak demand (see Fig. 10.9b) in the evening hours. In contrast, the proposed building energy management shifts part of the controllable loads (i.e. the demand of the laundry, see Fig. 10.9a) to an earlier hour in order reduce the demand in the evening. The lower generation of the non-dispatchable generation (see Fig. 10.9e) can be compensated by the grid connection (see Fig. 10.9g). In the evening, only the 10 kW diesel generator (see Fig. 10.9c) is required to cover the loads (see Fig. 10.9a, b) of the energy system.

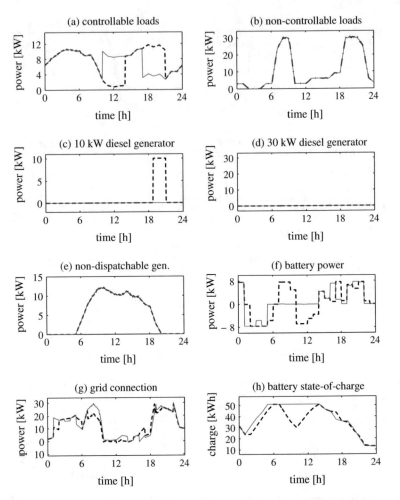

Fig. 10.8 Comparison of the simulation results obtained with the proposed building energy management (*solid line*) and the strategy without demand flexibility (*dash-dotted line*) for a typical demand profile of the considered hotel

The performances of the strategy without demand flexibility and the proposed building energy management after changes in the predictions are given in Fig. 10.10. It can be observed in the results that the strategy without demand flexibility uses between 7 am and 10 am the battery to compensate the lower non-dispatchable generation (see Fig. 10.9e). During the evening, both diesel generators (see Fig. 10.9c, d) are necessary to cover the loads (see Fig. 10.9a, b) of the energy system. The proposed building energy management easily compensates the reduced generation of the photovoltaic panels (see Fig. 10.9e) using the grid connection (see Fig. 10.9g) and the battery (see Fig. 10.9h). The energy management shifts part of the controllable load (see Fig. 10.9a) to an earlier hour to reduce the total load in the

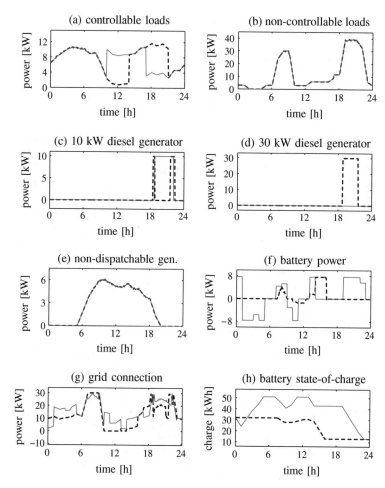

Fig. 10.9 Comparison of the simulation results obtained with the proposed building energy management (*solid line*) and the simple energy balancing strategy (*dash-dotted line*) for a typical demand profile with changes in the used predictions

evening hours. As a consequence, only the 10 kW diesel generator (see Fig. 10.9g) is necessary to cover the peak demand of the energy system.

The application of the energy management strategies to a hotel energy system with changes in the predictions led to considerably different results. Under consideration of the energy prices given for the different sources (see Sect. 10.4.1), the energy costs for the simple energy balancing strategy are 97.27 €, i.e. 0.229 €/kWh. In the case of the strategy without demand flexibility the energy costs of 89.19 € resulted in an energy price of 0.210 €/kWh. For the proposed building energy management the energy costs add up to 75.74 with an energy price of 0.178 €/kWh. The direct comparison shows that the proposed energy management reduced the

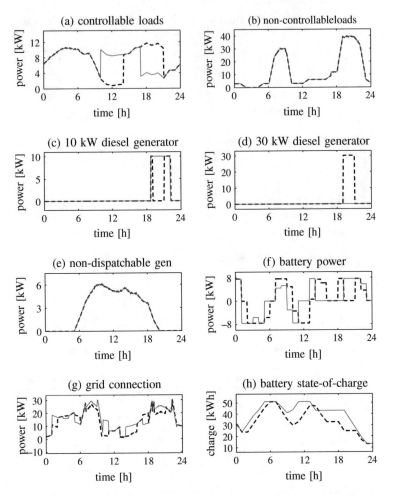

Fig. 10.10 Comparison of the simulation results obtained with the proposed building energy management (*solid line*) and the strategy without demand flexibility (*dash-dotted line*) for a typical demand profile with changes in the used predictions

energy price by 22.1 and 15.1% with respect to the simple energy balancing and the strategy without demand flexibility, respectively.

In the presented simulation the solutions to the short and medium term optimization problems of the building energy management were easily computed within the used sampling times. The calculation time for the medium term optimization ranged from $t^{m,min} = 0.328$ s and $t^{m,max} = 0.431$ s with an average of $t^{m,avg} = 0.355$ s. The short term optimization problem was solved in average in $t^{s,avg} = 0.357$ s, with a minimum of $t^{s,min} = 0.312$ s and a maximum of $t^{s,max} = 0.461$ s.

10.6   Conclusions

An advanced building energy management based on a receding horizon strategy has been developed. The strategy determines the optimal scheduling for all components of the local energy system, including energy storage, generation systems and controllable demands. The proposed two-stage approach solves two interconnected optimization problems—one for the medium term and another for the short term—considering the predictions of the local generation and consumption as well as the physical constraints of the energy system. The optimization problems include both continuous and discrete decision variables and can be expressed as mixed integer linear programs.

The optimal scheduling for the building energy system is computed in two successive steps. In the first place, the approach solves the medium term optimization problem over a horizon of one day with a sampling time of one hour. In the second place, the results obtained in the medium term optimization are used to define the short term optimization problem. The solution to the second problem contains the optimal scheduling for the energy system over a horizon of 1 h with a sampling time of 5 min. The proposed building energy management determines the optimal scheduling for a long horizon (one day) with a small sampling time (5 min) by solving two optimization problems of relatively low computational complexity.

The building energy management was implemented in the Matlab environment and additional toolboxes were used to solve the optimization problems. The performance of the proposed approach was applied to the simulated energy system of a medium size hotel. A case study with two other energy management strategies verified the good performance of the two-stage optimization. The proposed approach achieved a reduction between 15.1 and 29.4% with respect to the other energy management strategies used for comparison purposes. In the presented simulations the two-stage optimization calculated the optimal scheduling well within the used sampling time. The low computational burden allows using the building energy management on devices with reduced computation power such as smart meters or low-cost embedded systems.

The presented results illustrate the potential of advanced energy management in the building sector. The optimal scheduling of energy systems allows end-users to combine local generation systems, energy storage devices and demand response techniques and to minimize energy costs. Utilities and network operators benefit from a reduction of the effects of intermittent production and demand variations.

References

1. Iwaro, J., Mwasha, A.: A review of building energy regulation and policy for energy conservation in developing countries. Energy Policy **38**(12), 7744–7755 (2010)
2. Pérez-Lombard, L., Ortiz, J., Pout, C.: A review on buildings energy consumption information. Energy Build. **40**(3), 394–398 (Mar 2008)

3. dos Santos, A.H.C., Fagá, M.T.W., dos Santos, E.M.: The risks of an energy efficiency policy for buildings based solely on the consumption evaluation of final energy. Int. J. Electr. Power Energy Syst. **44**(1), 70–77 (2013)
4. Finn, P., O'Connell, M., Fitzpatrick, C.: Demand side management of a domestic dishwasher: Wind energy gains, financial savings and peak-time load reduction. Appl. Energy **101**, 678–685 (2013)
5. Moura, P.S., de Almeida, A.T.: The role of demand-side management in the grid integration of wind power. Appl. Energy **87**(8), 2581–2588 (2010)
6. Mohsenian-Rad, A.H., Leon-Garcia, A.: Optimal residential load control with price prediction in real-time electricity pricing environments. IEEE Trans. Smart Grid **1**(2), 120–133 (2010)
7. Palensky, P., Dietrich, D.: Demand side management: Demand response, intelligent energy systems, and smart loads. IEEE Trans. Industr. Inf. **7**(3), 381–388 (2011)
8. Tiptipakorn, S., Lee, W.J.: A residential consumer-centered load control strategy in real-time electricity pricing environment. In: Proceedings of the 39th North American Power Symposium (NAPS '07). pp. 505–510. Las Cruces, NM (30 Sept–2 Oct 2007)
9. Chen, C., Wang, J., Heo, Y., Kishore, S.: MPC-based appliance scheduling for residential building energy management controller. IEEE Trans. Smart Grid **4**(3), 1401–1410 (2013)
10. Scherer, H.F., Pasamontes, M., Guzmán, J.L., Álvarez, J.D., Camponogara, E., Normey-Rico, J.E.: Efficient building energy management using distributed model predictive control. J. Process Control. **24**(6), 740–749 (2014)
11. Asare-Bediako, B., Kling, W.L., Ribeiro, P.F.: Multi-agent system architecture for smart home energy management and optimization. In: Proceedings of the 4th IEEE PES Innovative Smart Grid Technologies Europe (ISGT Europe). pp. 1–5. Lyngby, Denmark (6–9 Oct 2013)
12. Zhao, P., Suryanarayanan, S., Simoes, M.G.: An energy management system for building structures using a multi-agent decision-making control methodology. IEEE Trans. Ind. Appl. **49**(1), 322–330 (Jan–Feb 2013)
13. Malysz, P., Sirouspour, S., Emadi, A.: MILP-based rolling horizon control for microgrids with battery storage. In: Proceedings of the 39th Annual Conference of the IEEE Industrial Electronics Society (IECON 2013). pp. 2099–2104. Vienna, Austria (10–13 Nov 2013)
14. Téllez Molina, M.B., Gafurov, T., Prodanovic, M.: Proactive control for energy systems in smart buildings. In: Proceedings of the 2011 2nd IEEE PES International Conference and Exhibition on Innovative Smart Grid Technologies (ISGT Europe). pp. 1–8. Manchester, UK (5–7 Dec 2011)
15. Atzeni, I., Ordőez, L.G., Scutari, G., Palomar, D.P., Fonollosa, J.R.: Noncooperative and cooperative optimization of distributed energy generation and storage in the demand-side of the smart grid. IEEE Trans. Signal Process. **61**(10), 2454–2472 (2013)
16. Tarasak, P., Chai, C.C., Kwok, Y.S., Oh, S.W.: Demand bidding program and its application in hotel energy management. IEEE Trans. Smart Grid **5**(2), 821–828 (2014)
17. Costanzo, G.T., Zhu, G., Anjos, M.F., Savard, G.: A system architecture for autonomous demand side load management in smart buildings. IEEE Trans. Smart Grid **3**(4), 2157–2165 (2012)
18. Zhao, Z., Lee, W.C., Shin, Y., Song, K.B.: An optimal power scheduling method for demand response in home energy management system. IEEE Trans. Smart Grid **4**(3), 1391–1400 (2013)
19. He, M., Murugesan, S., Zhang, J.: A multi-timescale scheduling approach for stochastic reliability in smart grids with wind generation and opportunistic demand. IEEE Trans. Smart Grid **4**(1), 521–529 (2013)
20. Moradzadeh, B., Tomsovic, K.: Two-stage residential energy management considering network operational constraints. IEEE Trans. Smart Grid **4**(4), 2339–2346 (2013)
21. Danandeh, A., Zhao, L., Zeng, B.: Job scheduling with uncertain local generation in smart buildings: Two-stage robust approach. IEEE Trans. Smart Grid **5**(5), 2273–2282 (2014)
22. Li, D., Jayaweera, S.K., Naseri, A.: Auctioning game based demand response scheduling in smart grid. In: Proceedings of the 2011 IEEE Online Conference on Green Communications (GreenCom). pp. 58–63. New York, NY (26–29 Sept 2011)

23. Qian, L.P., Zhang, Y.J.A., Huang, J., Wu, Y.: Demand response management via real-time electricity price control in smart grids. IEEE J. Sel. Areas Commun. **31**(7), 1268–1280 (2013)
24. Hillier, F., Lieberman, G.: Introduction to operations research. McGraw-Hill, New York, seventh edn. (2001)
25. Schrijver, A.: Theory of Linear and Integer Programming. Wiley, Chichester (1998)
26. Camacho, E.F., Bordons, C.: Model Predictive Control. Springer, London (2004)
27. Maciejowski, J.M.: Predictive Control with Constraints. Prentice Hall, Essex (2002)
28. Rossiter, J.A.: Model-Based Predictive Control: A Practical Approach. CRC Press, Boca Raton (2003)
29. Lofberg, J.: YALMIP: a toolbox for modeling and optimization in MATLAB. In: Proceedings of the 2004 IEEE International Symposium on Computer Aided Control Systems Design. pp. 284–289. Taipei, Taiwan (4 Sept 2004)
30. Achterberg, T.: SCIP: solving constraint integer programs. Math. Program. Comput. **1**(1), 1–41 (2009)
31. Achterberg, T., Berthold, T., Koch, T., Wolter, K.: Constraint integer program-ming: A new approach to integrate cp and mip. In: Perron, L., Trick, M.A. (eds.) Integration of AI and OR Techniques in Constraint Programming for Combinatorial Optimization Problems. Lecture Notes in Computer Science, vol. 5015, pp. 6–20. Springer, Berlin Heidelberg, Berlin (2008)
32. Currie, J., Wilson, D.I.: OPTI: Lowering the barrier between open source opti- mizers and the industrial MATLAB user. In: Proceedings of the 2012 Foundations of Computer-Aided Process Operations (FOCAPO). pp. 1–6. Savannah, GA (8–11 Jan 2012)

Chapter 11
An Investigation of Indoor Air Quality in UK Passivhaus Dwellings

Gráinne McGill, Tim Sharpe, Lukumon Oyedele, Greg Keeffe and Keith McAllister

Abstract The adoption of the German Passivhaus Standard in the UK has grown rapidly in recent years. Stimulated by the shift towards energy efficient design and rising fuel costs, the concept is perceived as a potential means of meeting energy and carbon targets through an established, reliable methodology. However the performance of the Standard in terms of adequate indoor air quality and thermal comfort in a UK climate remains under-researched. This paper describes the use of the Passivhaus Standard in a UK context, and its potential implications on indoor environmental quality. A case study is presented, which included indoor air quality measurements, occupant diary, building survey and occupant interviews in a Passivhaus social housing project in Northern Ireland. The study found issues with indoor air quality, the use and maintenance of Mechanical Ventilation with Heat Recovery (MVHR) systems, lack of occupant knowledge and the perception of overheating in the case study dwellings. The findings provide a much needed insight into the indoor environmental quality in homes designed to the Passivhaus standard; which can be disseminated to aid the development of an effective sus-

G. McGill (✉) · T. Sharpe
Mackintosh Environmental Architecture Research Unit,
Glasgow School of Art, Glasgow, UK
e-mail: g.mcgill@gsa.ac.uk

T. Sharpe
e-mail: t.sharpe@gsa.ac.uk

L. Oyedele
Bristol Enterprise Research and Innovation Centre,
University of West of England, Bristol, UK
e-mail: l.oyedele@uwe.ac.uk

G. Keeffe · K. McAllister
School of Planning, Architecture and Civil Engineering,
Queen's University Belfast, Belfast, UK
e-mail: g.keeffe@qub.ac.uk

K. McAllister
e-mail: k.mcallister@qub.ac.uk

© Springer International Publishing AG 2017
J. Littlewood et al. (eds.), *Smart Energy Control Systems for Sustainable Buildings*,
Smart Innovation, Systems and Technologies 67,
DOI 10.1007/978-3-319-52076-6_11

245

tainable building design that is both appropriate to localised climatic conditions and also sensitive to the health of building occupants.

Keywords Passivhaus · Indoor air quality · Overheating · Energy efficiency

11.1 Introduction

In order to meet the legally binding Climate Change Act (2008) of an 80% reduction of net carbon account by 2050 of the 1990 baseline [1], the UK Government has set a target of 'zero carbon' for all new buildings including housing by 2016 [2]. In 2009, the Government revised the definition of zero carbon by introducing the concept of 'allowable solutions' to compensate for the most challenging reductions of carbon emissions on site [3]. In the 2011 budget document released by the UK Government entitled 'The Plan for Growth', standards were relaxed to remove unregulated emissions from the definition [4]. Despite this, the challenge of an 80% carbon emissions reduction by 2050 remains eminent.

The built environment is responsible for approximately 36% of Greenhouse Gas (GHG) emissions for the whole of the UK, with domestic operational carbon emissions 54% of the built environment total [5]. In response, a number of energy efficient design strategies have been implemented in the UK housing sector, including adoption of the German Passivhaus standard. These strategies aim to reduce building carbon dioxide emissions through increased fabric energy efficiency and the adoption of low carbon technologies.

The Passivhaus concept is a voluntary construction standard established in Germany by Professor Wolfgang Feist during the early 1990s [6]. In the UK, adoption of the Passivhaus standard remains in its relatively early stages with approximately 200 completed projects [7], despite over 37,000 Passivhaus certified buildings worldwide [8]. The standard requires adherence to specific criteria; most notably annual maximum space heating requirements of 15 kWh/m^2, maximum annual primary energy of 120 kWh/m^2, utilisation of Mechanical Ventilation with Heat Recovery (MVHR) and an air tightness (n50) of less than 0.6 h^{-1} [9].

Proposals have been made for the Passivhaus standard or similar stringent nonresidential standard to be utilised as mandatory requirements for all new buildings by the European Commission [10, 11]. However, questions remain concerning the applicability of the Passivhaus standard in the UK in which there are key differences, for example, climate, space standards and procurement. The effect of these measures on indoor air quality (IAQ) and thermal comfort remain unknown, particularly in a social housing context. Accordingly, this study aims to (1) investigate the IAQ and thermal comfort of Passivhaus social housing during summer and winter seasons (both physical and perceived), (2) explore the effect of occupant activities on IAQ, and (3) examine occupant knowledge and engagement of the specialist ventilation systems installed in these homes.

11.2 Background

The effect of energy efficient design strategies on occupant health and wellbeing remains significantly under-researched, despite emerging evidence suggesting a significant lack of skills and knowledge in the area. For instance, as discussed by Sullivan et al. [12], limited published studies of IAQ in nearly zero energy homes have been identified in the UK. This is supported by Femenias [13], who explained that demonstration projects for sustainable buildings are rarely monitored adequately, leading to insufficient learning being applied to future projects. As recommended by the Zero Carbon Hub Ventilation and IAQ Task Group, 'further research should be undertaken by [UK] Government to inform future amendments to Building Regulations guidance and ensure public health and safety' [14].

This has been implemented through the Technology Strategy Board's *Building Performance Evaluation* competition, which dedicated £8 m of funding for the performance evaluation of new build/refurbishment projects over four years (2010–2014). Initial findings indicate IAQ concerns in bedrooms of contemporary dwellings [15], with particular issues in relation to the provision of adequate ventilation [16]. Similarly, apprehensions have been expressed regarding overheating in Passivhaus dwellings and significant discrepancies have been observed between measured and predicted indoor temperatures using Passive House Planning Package software [17].

Correspondingly, emerging research from Europe suggests conflicting evidence on the effect on IAQ and thermal comfort in Passivhaus dwellings. On one hand, a review of post occupancy evaluation studies in passive houses in Central Europe by Mlecnik et al. [18] found users of passive houses usually feel more comfortable in winter months compared to summer months. They suggest that further attention to overheating is required in order to improve user satisfaction. Issues with perception of IAQ and knowledge of the heating and ventilation equipment were also highlighted. This criticism of the Passivhaus Standard is supported by McLeod [19], who states that Passivhaus dwellings are inherently vulnerable to overheating and suggests, 'active cooling systems may become a de facto requirement in urban Passivhaus and low energy dwellings in the UK within the next 30–40 years.' Emerging health risks associated with passive houses were also highlighted in studies by Hasselaar and Hens [20, 21].

Conversely, a number of studies have suggested improved IAQ and thermal comfort [22–24] in Passivhaus dwellings. For instance, Mlecnik et al. [25] refer to a study where occupants of Passivhaus homes reported improved freshness of air in the bedrooms during the morning period. However, the suitability of the Passivhaus standard for the UK context remains a contentious issue, with questions regarding the necessity and/or desirability of MVHR in UK dwellings [26–28], particularly in a social housing context. This study therefore seeks to examine the IAQ and thermal comfort in UK Passivhaus social housing through a case study investigation.

11.3   Methodology

The three selected case study dwellings are within a block of five built to the Passivhaus standard, located in a residential area in Northern Ireland. Three mid-terraced Passivhaus dwellings were selected for investigation following discussions with the Housing Association[1] and building occupants. The two-storey, 3–4 bedroom timber frame dwellings also achieved Level 4 in the Code for Sustainable Homes and are compliant with the Lifetime Homes Standard. The development is south facing with main entrance and car-parking to the north (Table 11.1).

Occupant interviews and building surveys were conducted in the three Passivhaus case study dwellings to help obtain information about perception of IAQ, thermal comfort, sick building syndrome (SBS) symptoms and general building conditions. IAQ measurements were conducted in all three dwellings during the summer (May 2013) and in two of the three dwellings during the winter season (Feb–March 2013). Occupant diaries completed during the periods when air quality measurements were taken helped to provide information on occupancy levels, heating schedules and activities that may have affected the results. Construction of the dwellings was completed in April 2012.

11.3.1   Indoor Air Quality Measurements

The IAQ measurements were conducted for approximately 24 h in the ground floor open plan living room and kitchen of each dwelling during the summer and winter season (2013). The living area is south facing, opening onto an external shaded patio and rear garden. The façade consists of large, triple glazed doors and fixed glazing with brise soleil for shading. There is a double height glazed section over the dining area, with fixed shutters for shading (see Fig. 11.1). Measurement parameters include temperature, dew point, wet bulb, relative humidity, carbon dioxide (monitored with Extech EA80 Datalogger/accuracy ±3%, ±5 °C, ±3% of reading or ±50 ppm) and formaldehyde (monitored with HalTech HFX205/accuracy ±5%). Measurements were also conducted in the main bedroom during summer months (monitored with Wohlër CO_2 datalogger/ accuracy ±3% RH, ±0.6 °C, ±3% of reading or ±50 ppm). All IAQ measurements were conducted in accordance with ISO 16000. Outside conditions were monitored during the measurement period with use of a weather station (Watson W-8681-SOLAR) and Wohlër CO_2 datalogger.

Measurements of Volatile Organic Compounds (VOCs) were conducted in house No. 1 during the winter season (2014). Air samples were collected simultaneously in the kitchen, living room, main bedroom and outside. Indoor samplers were positioned at breathing height and away from possible sources of pollution.

---

[1]Housing Associations are voluntary organisations that aim to help people to acquire affordable accommodation that meet their requirements.

Table 11.1 Household profiles

Household profiles	No. 1	No. 2	No. 3
No. of occupants	4	6	3
Cooking fuel	Electric	Electric	Electric
Heating fuel	Natural gas	Natural gas	Natural gas
No. of smokers	1	3	1
Cigarettes ever smoked in home	No	No	No
Average hours occupied during week	22	24	24
Average hours occupied at weekend	24	24	22

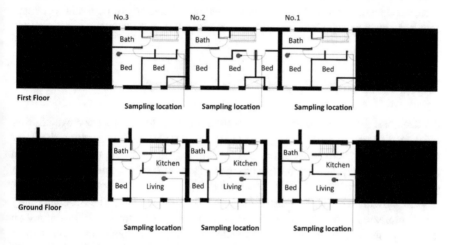

Fig. 11.1 Floor plan and sampling location

The outdoor sampler was positioned in the back garden of house No. 1, away from exhaust vents, openings and direct solar irradiation. Two field blank samples were also taken. A pumped sampling method was deployed (as described in ISO 16017-1:2000) where sorbent tubes packed with an adsorbent (HEYSEP—packed in house) were connected to a pump (pocket pump, SKC, Dorset) at a flow rate of 100 ml/min for approximately 2 h. The collected vapour on each sampling tube was then desorbed using a thermal desorber (ATB 400, PerkinElmer, Cambridge) and transferred into a gas chromatograph (Turbomass GC Mass Spec, PerkinElmer, Cambridge) equipped with a mass spectrometer and a RTX 5 capillary column (50 m, 0.25 mm). Analytical calibration was achieved through liquid spiking onto a sorbent tube. All quantification was achieved relative to toluene. Outside values for hexanal, ethanol and terpenes were not reported as concentrations were below the detectable limit.

11.3.2 Structured Occupant Interviews

Structured interviews were conducted with building occupants of all three Passivhaus dwellings, utilising specifically composed questionnaires. The questionnaires consisted of one for each household, one for each occupant (adults) and one for each child (to be completed by a parent/guardian). The interviews obtained information on the perception of IAQ and thermal comfort during summer and winter seasons, in addition to the presence of any Sick Building Syndrome symptoms (SBS) and Building Related Illnesses (BRI), utilising validated procedures [29, 30]. The household questionnaire gathered information on the building occupants, ventilation strategies, building features, frequency of particular occupant activities, heating schedules and the use of air polluting products.

11.3.3 Occupant Diary and Building Survey

An occupant diary was used to gain information on activities that occurred during the measurement period that might have affected the results. For instance, the diary required the occupants to record average occupancy in the living room/kitchen and in the home every hour. Hourly activities (such as heating, cooking, use of air polluting products, opening of doors/windows, and use of boost mode in the MVHR system) were also recorded through a tick-box method. The occupant diary was compressed to one A4 page for each measurement day, to reduce the burden on the occupants. The building survey recorded information on building features, such as the presence of operable windows, floor coverings and general observations, in addition to heating and ventilation controls. The survey was conducted on the day of the measurements.

11.4 Results

11.4.1 Heating and Ventilation

The three households were asked a number of questions about operation, maintenance and general knowledge of the Mechanical Ventilation with Heat Recovery (MVHR) system. The results suggest significant issues that require attention, particularly in a social housing context. Specifically, all three households were asked if they have ever had any issues with the MVHR system since they moved in. Two of the households stated that the MVHR system had broken down and was now not working; one stated it had broken down a month before the interviews (No. 2) and the other eight months before the interviews (No. 3). The occupants of No. 3 then went on to explain that there was a problem with the electrics and there appeared to be difficulty finding people in the local area with adequate expertise of MVHR systems.

Knowledge of the ventilation system was also an issue in the case study dwellings. For example, only one household was aware of the boost mode function and used it regularly (No. 3). Furthermore, when asked if they had ever adjusted the supply or extract vents, an occupant of house No. 3 stated; *'yes, I usually have them wider open; it doesn't affect how much air coming in, it affects the noise. (…) I close them in the bathroom sometimes because when my son gets bathed when we keep it open it extracts the air and it is cooling him as well.'* Thus the importance of balancing the MVHR system and the impact of adjusting the vents was not clearly understood by the building tenant.

All occupants had been living in the home for approximately one year when interviewed. When asked if the filters in the ventilation system had been replaced, one household stated they were not sure since maintenance of the MVHR system is the responsibility of the Housing Association. The other two households stated that although the filters needed changed, it had not yet been done. Afterwards when informed of this fact, the Housing Association explained that access to the properties for maintenance of the MVHR systems had been problematic. As a result, they have now decided to schedule the maintenance of the MVHR system in the future to coincide with the annual boiler servicing, since access for boiler servicing is a legal requirement in social housing. Households were also asked if they have ever had any issues with the MVHR system (such as noise, cost of running, thermal comfort, draughts or other). Household No. 1 stated *'yes, the system is noisy on higher settings.'* Household No. 3 also indicated that they had experienced problems with both thermal comfort and draughts in their home.

Households were asked how often the windows were opened during the summer and winter months; the results of which are illustrated in Fig. 11.2. All households reported opening the window either 'regularly' or 'constantly' in the morning, during the day and at night, during the summer months. Two households (No. 1 and No. 3)

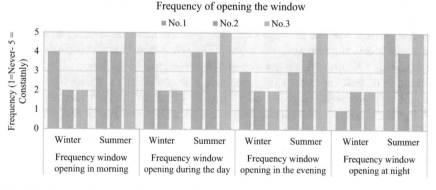

Fig. 11.2 Household reported frequency of opening windows during summer and winter season (*1* Never, *2* Rarely, *3* Occasionally, *4* Regularly, *5* Constantly)

Table 11.2 Heating schedule

Heating schedule	No. 1	No. 2	No. 3
Winter	6–7 pm	10–12 am	5–8 pm
Spring	–	5–8 pm	–
Summer	–	5–6.30 pm	–
Autumn	6–7 pm	1–4 pm	5–7 pm

explained that it was too warm indoors. During the winter months, occupants reported opening the windows less, with two households 'rarely' opening the window at any time of day (No. 2 and No. 3). These two households however also reported that the MVHR system was not working, which may lead to significant problems with ventilation during the winter months.

The homes are heated by one radiator in the lounge and two towel radiators in the bathrooms. A post heater (hot water heating coil connected to the thermal store) is also available in the MVHR system, which is controlled by a thermostat in the entrance hallway. Table 11.2 illustrates the reported heating schedule of each household for each season. Household No. 1 and No. 3 reported using the central heating system for approximately 1–3 h in the evening during autumn and winter. In house No. 2 however, the heating system was used regularly during all seasons.

11.4.2   Carbon Dioxide and Average Occupancy in Open Plan Living Area

In these studies carbon dioxide (CO_2) is being used as an indicator of ventilation rates. Levels of CO_2 correlate well with human occupancy and human-generated pollutants, but may be unconnected from pollutants not related to occupancy, such as off-gassing from building materials, carpets and furniture. Nevertheless, in the context of concern over ventilation rates, they provide a useful indicator of relative levels of ventilation. There is a general acceptance that CO_2 keeps 'bad company' and that levels above 1000 ppm are indicative of poor ventilation rates [31], which corresponds to a ventilation rate of 8 l/s per person [32]. This figure is also relevant in comparison with the findings of a review of the literature looking at the associations between ventilation rates and CO_2 levels with health outcomes which concluded, "Almost all studies found that ventilation rates below 10 Ls^{-1} per person in all building types were associated with statistically significant worsening in one or more health or perceived air quality outcomes" [33]. A recent paper by Wargocki identified associations between CO_2 levels and health and concluded "The ventilation rates above 0.4 h^{-1} or CO_2 below 900 ppm in homes seem to be the minimum level to protect against health risks based on the studies reported in the scientific literature" [34].

The CO_2 level of 1000 ppm [35] was exceeded in both the two measured households during the winter (Figs. 11.3 and 11.4) and all three households during the summer measurement period (Figs. 11.5, 11.6 and 11.7). Levels peaked as high

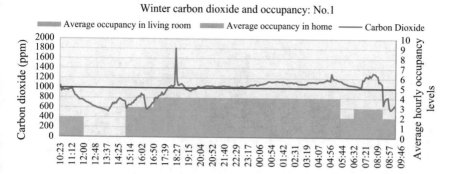

Fig. 11.3 Winter living space carbon dioxide and occupancy in House No. 1

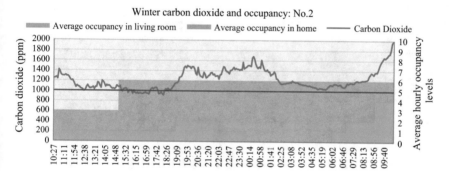

Fig. 11.4 Winter living space carbon dioxide and occupancy in House No. 2

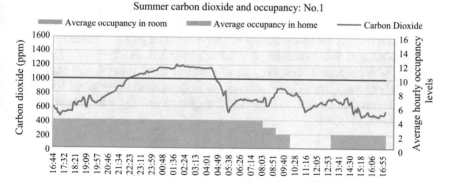

Fig. 11.5 Summer living space carbon dioxide and occupancy in House No. 1

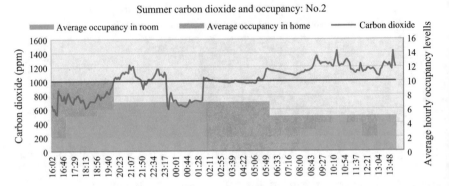

Fig. 11.6 Summer carbon dioxide levels in House No. 2

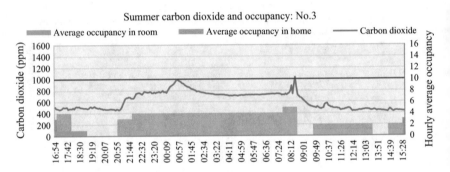

Fig. 11.7 Summer carbon dioxide levels in House No. 3

as 1992 ppm during winter measurements in House No. 2. Mean CO_2 levels also exceeded 1000 ppm during both summer and winter months in house No. 2.

In House No. 1, CO_2 levels did not correspond with reported occupancy levels in the room during the measurement period. For instance, during the early hours of the morning when reported occupancy in the measurement room (open plan living room/kitchen) was zero, CO_2 levels remained high; in most cases above 1000 ppm. This might possibly be due to air leakage from the conjoining bedrooms above into the double height space.

11.4.3   Summer Bedroom Conditions

Bedroom carbon dioxide, temperature and relative humidity were recorded during the summer months in all three households, as illustrated in Table 11.3. Carbon dioxide levels varied significantly during the measurement period, with peak levels ranging from 804 to 2598 ppm. Carbon dioxide levels in House No. 3 were significantly high

Table 11.3 Summer bedroom carbon dioxide, temperature and relative humidity

	CO_2 (ppm)			Temp (°C)			RH (%)			V.P.
	Max	Min	Mean	Max	Min	Mean	Max	Min	Mean	Mean
No. 1	804	407	590.0	25.4	20.2	22.6	51.8	35.3	42.7	1.17
No. 2	1520	436	782.8	22.5	19.1	21.1	53.0	36.8	45.3	1.13
No. 3	2598	396	820.3	22.6	19.6	21.5	47.7	31.9	41.3	1.06

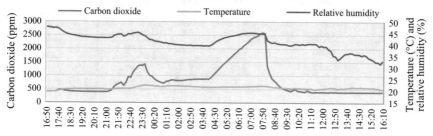

Summer bedroom conditions: No. 3

Fig. 11.8 Summer carbon dioxide, temperature and relative humidity in House No. 3

through the night, suggesting inadequate ventilation at this time. Levels dropped significantly between 8 and 9 am, most likely as a result of purge ventilation in the form of opening window(s) (Fig. 11.8). Recorded temperatures in house No. 1 peaked above the CIBSE's 'hot' temperature threshold for bedrooms of 25 °C [36], indicating problems with overheating Table 11.3.

11.4.4 Living Room Relative Humidity and Temperature

Relative humidity levels remained below the recommended maximum of 60% in all households during both summer and winter measurements (see Table 11.4). No significant difference between summer and winter seasons was found. However, as illustrated in Figs. 11.9 and 11.10, an examination of actual vapour pressure levels identified areas of concern regarding threshold levels for dust mite control. Specifically, vapour pressure levels exceeded 1.13 kPa (or 7 g of water vapour per kg of dry air), in all dwellings during both summer and winter measurements. As explained by Korsgaard and Harving [39, 40] the recommended maximum vapour pressure level of 1.13 kPa (or 7 g/kg) corresponds to a Threshold Limit Value (TLV) for house dust mite exposure of 100 mites/g of dust. This value has been derived from the literature as an exposure level 'below which no increased disease frequency can be associated with the actual exposure' [39] (p. 78). This is supported by Platt-mills et al., who stated that maintaining absolute humidity levels below

Table 11.4 Statistical analysis of relative humidity and temperature

Parameter	Statistical analysis	No. 1		No. 2		No. 3
		Summer	Winter	Summer	Winter	Summer
Relative humidity	Maximum	54.0	52.4	51.0	51.0	57.9
	Minimum	33.6	37.0	38.8	42.5	34.4
	Mean	43.4	43.1	45.7	46.3	46.9
	Standard Dev.	4.7	2.6	2.7	1.8	4.1
Temperature	Maximum	24.9	24.6	24.0	23.2	23.3
	Minimum	19.0	19.0	20.5	19.7	18.9
	Mean	23.2	22.2	22.4	21.5	21.0
	Standard Dev.	1.2	0.9	0.7	1.0	0.9
Vapour pressure	Maximum	1.70	1.62	1.52	1.45	1.66
	Minimum	0.74	0.81	0.94	0.97	0.75
	Mean	1.23	1.15	1.24	1.19	1.17

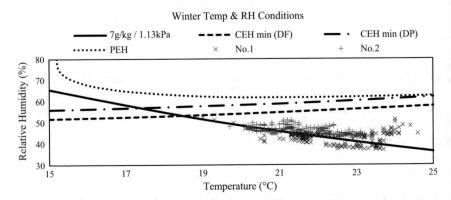

Fig. 11.9 Winter living space temperature and relative humidity. (CEH (DP) refers to Critical Equilibrium Humidity for *Dermatophagoides pteronyssinus* [37], a dust mite species common to the UK. CEH (DF) is the Critical Equilibrium Humidity for *Dermatophagoides farinae* [38], common dust mite species of USA. PEH refers to Population Equilibrium Humidity.)

7 g/kg (1.13 kPa) should reduce the risk of excess mite growth [41, 42] which may be considered significant at lower temperatures (15°–18°).

The Critical Equilibrium Humidity (CEH) for *Dermatophagoides farinae* (DF) was exceeded during the summer measurements in House No. 3; however conditions remained below the CEH for *Dermatophagoides pteronyssinus* (DP) during both summer and winter measurements. It is suggested therefore that relatively high vapour pressure levels are being masked to a degree by higher indoor temperatures in the case study dwellings. Average temperatures remained within satisfactory levels for comfort (18–24 °C) [43, 44] during the measurement periods, ranging from 21 to 23.2 °C, however peaked above 24 °C in house No. 2 during both summer and winter months.

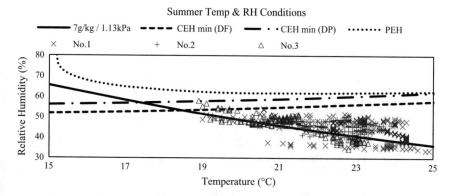

Fig. 11.10 Summer living space temperature and relative humidity

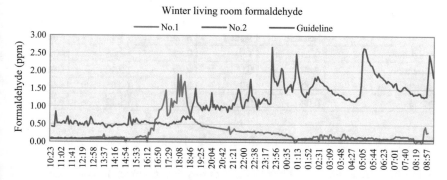

Fig. 11.11 Winter living space formaldehyde levels

11.4.5 *Formaldehyde Levels in Open Plan Living Room and Kitchen*

Peak levels of formaldehyde in the open plan living room and kitchen in two households (No. 1 and No. 2) significantly exceeded the WHO recommended 30 min time weighted average of 0.08 ppm [45]. In House No. 2, winter levels peaked at 2.68 ppm; over 30 times the recommended 30 min average. Mean values over the measurement period exceeded 0.08 ppm during the summer and winter of House No. 2 and during the winter in House No. 1. Winter and summer levels over the measurement period are illustrated in Figs. 11.11 and 11.12.

During the winter period of House No. 1 where the levels of formaldehyde peaked significantly (4–8 pm), the occupants reported drying clothes naturally indoors (3–11 pm) and the use of the cooker (3–5 pm). In house No. 2, the levels of formaldehyde were significantly high throughout the winter measurement period, and do not seem to correspond with activities recorded through use of the occupant

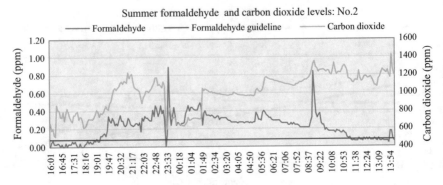

Fig. 11.12 Summer living space formaldehyde and carbon dioxide levels in House No. 2

Table 11.5 Statistical analysis of formaldehyde levels in open plan living room and kitchen

	No. 1		No. 2		No. 3
	Summer	Winter	Summer	Winter	Summer
Maximum	0.23	1.91	0.89	2.68	0.08
Minimum	0.00	0.00	0.00	0.40	0.00
Standard deviation	0.02	0.32	0.14	0.55	0.01
Mean	0.00	0.28	0.22	1.13	0.01

diary. Three occupant's in House No. 2 smoke which may have contributed to the high results, despite occupants stating that cigarettes were not smoked indoors.

Formaldehyde levels in House No. 1 during the summer measurement period peaked between 10 and 11 am. Occupants reported the use of cleaning products (8–10 am), opening external doors (9–10 am) and the use of the cooker (9–10 am) around this time, which may have contributed to the results. In House No. 3, summer formaldehyde levels recorded during the measurement period did not exceed 0.08 ppm (Table 11.5).

However, in House No. 2, formaldehyde levels were significantly high throughout the summer measurement period, with levels above 0.08 ppm between 19:16 on day one until 11:18 on day two, with a drop to 0.00 ppm from 23:23 to 23:33. During this time, the occupants recorded drying clothes naturally indoors (7 pm–6 am, 7–10 am) and the use of plug in air fresheners (4–2 pm). External doors were opened in the home from 7 to 8 pm, 9–10 pm and 6–8 am (Table 11.6). In addition, as presented in Fig. 11.12, a loose relationship existed between the carbon dioxide and formaldehyde levels over the 24 h measurement period. This supports the premise of using carbon dioxide levels as an indicator of poor ventilation; however as illustrated, is not conclusive.

Table 11.6 Frequency of activities during the measurement period (obtained by occupant diary)

Reported activities	No. 1		No. 2		No. 3
	Summer	Winter	Summer	Winter	Summer
Use of air fresheners	–	9–10 am, 6–7 am	4 pm–2 pm	10 am– 1 pm, 4– 8 pm	10–11 am
Drying clothes naturally indoors	–	3–11 pm	7 pm–6 am, 7–10 am	–	10 am– 5 pm
Use of cleaning products	8–10 am	2–3 pm	–	8–9 am, 2–3 pm, 8–10 pm	4–5 pm
Use of incense/scented candles	–	–	–	–	–
Use of boost mode in MVHR	–	–	–	–	–
Use of the cooker	5–6 pm, 8–9 pm 9–10 am, 1–2 pm	3–5 pm, 6–7 am	4–5 pm, 7–8 am	8–9 am, 2–3 pm, 8–10 pm	4–5 pm
External doors opened	5–8 pm, 9–10 am	10–11 am, 2–3 pm, 8–9 pm	4–6 pm, 7–8 pm, 9–10 pm, 6–8 am, 10–11 am	–	4–5 pm, 10 am– 5 pm

11.4.6 Volatile Organic Compounds (VOC's) in House No. 1

In addition to the environmental monitoring, measurements of VOCs were conducted in House No. 1 in the living room, kitchen, bedroom and outside during the winter season, as illustrated in Table 11.7. In general, concentrations of VOCs did not vary significantly from room to room. However, concentrations of xylenes were significantly higher in the kitchen of House No. 1 (16.98 $\mu g/m^3$) in comparison to the living room (4.24 $\mu g/m^3$) and bedroom (3.70 $\mu g/m^3$). This may be due to the fact that the kitchen is north facing overlooking the main road, thus outdoor sources of xylenes from exhaust fumes may have influenced the results. Furthermore, indoor concentrations of all measured VOC's were significantly higher than outdoor concentrations; with the exception of benzene.

Concentrations of all measured VOCs were below recommended maximum levels. However, benzene and styrene have both been classified by the Environmental Protection Agency (EPA) as known or possible human carcinogens; thus exposure should be limited in a domestic environment. In comparison, a study of 876 homes in England (selected using the Survey of English Housing) found higher levels of benzene (mean = 3.0 $\mu g/m^3$) indoors [46]. However, mean levels of hexanal were significantly lower (0.9 $\mu g/m^3$), in comparison to the case study

Table 11.7 VOC results in living room, kitchen, bedroom and outside in house No. 1

VOCs	Living room (μg/m³)	Kitchen (μg/m³)	Bedroom (μg/m³)	Outside (μg/m³)	BRE survey [46, 47] (mean conc. μg/m³)
Terpene (No. 1)	101.73	125.20	77.46	n/a	n/a
Terpene (No. 2)	18.04	20.34	13.85	n/a	n/a
Terpene (No. 3)	43.27	39.77	34.19	n/a	n/a
Toluene	11.91	12.54	12.16	1.71	15.1
Ethyl benzene	2.83	2.71	2.39	0.81	1.2
Xylenes	4.24	16.98	3.70	1.55	3.8
Hexanal	31.34	25.21	29.15	n/a	0.9
Benzene	0.57	0.49	0.51	0.40	3.0
Cyclopentane	6.83	5.29	5.89	0.62	n/m
Ethanol	26.58	33.25	25.58	n/a	n/m
Pentane (No. 1)	56.78	43.48	47.22	1.10	n/a
Pentane (No. 2)	67.88	53.68	61.75	0.45	n/a
Styrene	3.73	3.37	2.92	0.10	n/m
TVOCs	330	300	280	<50	210

*n/m = not measured

results (25.2–31.3 μg/m³). TVOC concentrations exceeded the performance criteria set out in the UK building regulations[2] (Approved Document F: Ventilation 2010) of 300 μg/m³ in the open plan living area (330 μg/m³); with concentrations of 300 μg/m³ recorded in the kitchen and 280 μg/m³ in the main bedroom. Outside TVOC concentrations were <50 μg/m³.

11.4.7 Indoor Air Quality Perception

Occupant perception of IAQ was monitored through use of seven point uni-polar and bi-polar scales. For uni-polar scales such as 'fresh (1)—stuffy (7)' where one extreme is considered bad, a score greater than 3 requires further investigation and a scale greater than 5 is a cause for concern [23]. For bi-polar scales such as 'too still (1)—too draughty (7)' where both extremes are bad, scores outside 3–5 require further investigation and outside 2–6 is a cause for concern. The statistical analysis results were derived from results of all adult questionnaires utilised in the structured interviews for all three households.

As illustrated in Table 11.8, mean scores for the 'fresh stuffy' scale (3.5) and 'satisfactory overall unsatisfactory overall' scale (3.3) for the winter months requires further investigation. This suggests occupants of the case study dwellings

---

[2]Approved Document- Part F (Ventilation) 2010 for England and Wales (Part K in Northern Ireland) recommends performance criteria of 300 μg/m³ for TVOCs, averaged over 8 h.

Table 11.8 Perception of IAQ during winter in the Passivhaus households

IAQ perception scales	Mean	S.D	Mean + S.D	Mean − S.D	Max	Min
Dry(1)—humid(7)	3.3	0.5	3.8	2.8	4	3
Fresh(1)—stuffy(7)	3.5	1.3	4.8	2.2	5	2
Odourless(1)—odorous(7)	3.0	0.0	3.0	3.0	3	3
Too still(1)—too draughty(7)	3.0	1.4	4.4	1.6	4	1
Satisfactory overall(1)—unsatisfactory overall(7)	3.3	1.0	4.2	2.3	4	2

Table 11.9 Perception of IAQ during summer in the Passivhaus households

IAQ perception scales	Mean	S.D	Mean + S.D	Mean − S.D	Max	Min
Dry(1)—humid(7)	2.3	0.5	2.8	1.8	3	2
Fresh(1)—stuffy(7)	3.3	1.9	5.1	1.4	6	2
Odourless(1)—odorous(7)	3.0	1.4	4.4	1.6	5	2
Too still(1)—too draughty(7)	2.3	1.0	3.2	1.3	3	1
Satisfactory overall(1)—unsatisfactory overall(7)	3.8	2.2	6.0	1.5	7	2

did not perceive the air to be significantly fresh or satisfactory during winter. Similarly, during the summer, the mean score for the 'fresh stuffy' scale was 3.3 and the 'satisfactory overall unsatisfactory overall' scale was 3.8. Furthermore, mean scores for bipolar scales 'dry humid' and 'too still too draughty' were outside the range of 3–5, suggesting further investigation is required. It is important to note that the maximum score for the scale 'satisfactory overall unsatisfactory overall' during the summer month was 7, thus at least one occupant considered the IAQ as significantly unsatisfactory, which is certainly a cause for concern (Table 11.9).

11.4.8 Thermal Comfort Perception

Occupant perception of thermal comfort in the case study dwellings during the winter months was generally satisfactory, with all mean scores remaining within acceptable limits. However the minimum score for the scale 'too hot too cold' was 1, thus at least one occupant considered the home 'too hot' during the winter months (Table 11.10).

During the summer months, the mean score for the scale 'too hot too cold' was 1.8, which is a cause for concern and suggests problems with summertime overheating. Similarly, overall satisfaction of thermal comfort during the summer months requires further investigation, with an average score of 3.8. The maximum

Table 11.10 Perception of thermal comfort during winter

Thermal comfort perception scales	Mean	S.D	Mean + S.D	Mean − S.D	Max	Min
Comfortable(1)—uncomfortable(7)	2.0	0.0	2.0	2.0	2	2
Too hot(1)—too cold(7)	3.0	1.4	4.4	1.6	4	1
Stable(1)—varies throughout the day(7)	3.0	0.8	3.8	2.2	4	2
Satisfactory overall(1)—unsatisfactory overall(7)	2.8	0.5	3.3	2.3	3	2

Table 11.11 Perception of thermal comfort during summer

Thermal comfort perception scales	Mean	S.D	Mean + S.D	Mean − S.D	Max	Min
Comfortable(1)—uncomfortable(7)	3.0	2.0	5.0	1.0	6	2
Too hot(1)—too cold(7)	1.8	1.0	2.7	0.8	3	1
Stable(1)—varies throughout the day(7)	4.0	1.4	5.4	2.6	6	3
Satisfactory overall (1)—unsatisfactory overall (7)	3.8	1.5	5.3	2.3	6	3

score for the 'comfortable uncomfortable' scale was 6, which means at least one occupant considered the home as 'uncomfortable' during the summer season (Table 11.11).

11.5 Discussion

The results from the study suggest that there are significant issues with the effectiveness of the MVHR system in practice. These include: (1) design and installation issues; for example the importance of balancing the MVHR system, adjustment of the supply and extract vents and on-going system faults; (2) maintenance issues, such as lack of skilled service engineers and lack of filter replacements; and (3) occupant engagement, for example inadequate knowledge of the boost mode function, problems with noise on higher settings, draughts and problems with thermal comfort.

With regards to occupant engagement, during the handover stage the Housing Association provided a pre-allocation meeting with potential tenants, pre-handover viewings, user manuals and information posters in all dwellings. Since occupants were chosen based on a waiting list rather than environmental awareness and/or lifestyles, understanding and training was considered significantly important, particularly in a social housing context. The results however suggest that there are still improvements to be made to ensure adequate knowledge and understanding of the MVHR system from building occupants. In addition, it is suggested that a service checklist should be developed and implemented (at least once a year) to ensure adequate performance and maintenance of MVHR systems in social housing schemes.

During the summer months, occupants reported opening the windows either regularly or constantly in the morning, during the day and at night; with two households (No. 1 and No. 3) explaining it was too warm indoors. This suggests that the MVHR system alone was not capable of ensuring adequate thermal comfort. Both of these households reported using the central heating system for approximately 1 to 3 h a day during winter and autumn. Household No. 2 however stated that they utilised central heating regularly during all seasons. This suggests significant variances in heating schedules, which may have a major effect on the annual space heating demand. In winter, occupants reported opening the windows much less with two households 'rarely' opening windows at any time of day, which may cause problems where the MVHR systems are not performing adequately, but conversely, window opening in winter will undermine the effectiveness of the system for heat recovery.

The high levels of carbon dioxide (>1000 ppm) recorded in all monitored households during both summer and winter months suggest insufficient ventilation in the case study dwellings. This may be as a result of inadequate performance, use and/or maintenance of the MVHR system. Levels peaked as high as 2598 ppm in the bedroom of No. 3 during the summer measurement period. According to the German Working Group on Indoor Guideline Values, 'based on health and hygiene considerations: concentrations of indoor carbon dioxide below 1000 ppm are regarded as harmless, those between 1000 and 2000 ppm as elevated and those above 2000 ppm as unacceptable' [48]. In House No. 2, mean carbon dioxide levels exceeded 1000 ppm during both summer and winter months. More research is therefore required to investigate the performance of MVHR systems in practice and whether or not they are providing adequate ventilation in low-energy, Passivhaus dwellings; and whether heat recovery efficiencies are being undermined by adaptive behaviour to maintain comfortable conditions.

Levels of relative humidity remained reasonably low during both summer and winter months in monitored dwellings, with average values ranging from 43.1 to 46.9%, which may be partly due to the use of MVHR systems, with very little variance between summer and winter. However, an examination of vapour pressure illustrated the levels of moisture within the dwellings were high, exceeding 1.13 kPa in all dwellings and exceeding CEH (DP) in House No. 3 during the summer measurements. This suggests that the high temperatures and reasonably low relative humidity levels indoors may be disguising poor hydrothermal conditions in the case study dwellings.

With regards to thermal comfort in the dwellings, mean temperatures remained within satisfactory levels for comfort (18–24 °C) during both summer and winter measurements, however peaked above 24 °C in house No. 2. Despite this, occupants' general perception of thermal comfort was poor during the summer months, with perceived overheating a significant concern. Furthermore, at least one occupant in House No. 2 perceived the thermal comfort during winter months as 'too hot', suggesting problems with excessive internal sources of heat. This is supported by findings from a study of Passivhaus dwellings in Scotland, which reported similar issues with overheating, partially attributed to significant incidental heat gains through uninsulated hot water pipework identified through a thermography study [17]. In an effort to reduce energy demand in buildings through energy efficient strategies, architects must be careful to ensure potential savings are not offset through increased cooling requirements as a result of overheating. Awareness of this problem in new build, low energy dwellings is increasing through the publication of recent reports [19, 49–51].

Recorded formaldehyde levels over the monitoring period significantly exceeded the WHO 30 min time weighted average of 0.08 ppm, with winter levels reaching as high as 1.91 ppm (No. 1) and 2.68 ppm (No. 2). Winter levels were much higher than summer levels, possibly since occupants reported opening windows and/or external doors more frequently during the summer season, which would have helped dilute indoor concentrations (in turn illustrating the effectiveness of natural ventilation). Furthermore, homes had been occupied for longer during the summer measurements thus off-gassing from building materials would likely be reduced. The use of the occupant diaries suggested possible sources of formaldehyde from activities conducted during the measurement period. For instance, peaks in formaldehyde levels in House No. 1 (winter: 4 to 8 pm) and No. 2 (summer: 7 pm to 11 am) appeared to coincide with naturally drying of clothes indoors (No. 1: 3 to 11 pm; No. 2: 7 pm to 6 am; 7 to 10 am). This may be as a result of off-gassing of formaldehyde or VOCs from laundry products [52–54]. However House No. 3 reported drying clothes indoors during the measurement period and levels of formaldehyde did not exceed the recommended guideline of 0.08 ppm. The location of drying clothes indoors was not recorded, thus clothes may not have been dried in the measurement room. Measurements of VOCs in House No. 1 found indoor concentrations significantly higher than outside. VOC concentrations did not vary significantly between rooms, with the exception of xylenes, where higher levels were observed in the kitchen. All measured VOCs in House No. 1 were below recommended maximum levels.

Finally, the perception of IAQ recorded through the structured occupant interviews suggests occupants did not perceive the air quality to be significantly fresh or significantly satisfactory during summer or winter. Furthermore, mean scores suggest occupants perceived the air as relatively dry and still during the summer months, which may have implications on overall comfort. At least one occupant perceived the air quality as significantly unsatisfactory, which is a cause for concern. These results demonstrate convergence with the results of the IAQ measurements, and highlight the need for an urgent review of energy efficient design strategies and the effect on IAQ.

In particular, it is important to evaluate occupant knowledge and usability of mechanical systems, especially in the context of social housing. In theory, the

Passivhaus concept provides an established, systematic methodology supported by scientific literature to acquire the perfect performance, at least in terms of energy. However this must be envisaged in the presence of risk factors, such as occupant understanding, operation, and system performance, and the effect of these on overall performance. Moreover, exacerbating factors such as indoor pollutant concentrations, room volumes, and weather, play an important role in the resulting quality of the indoor environment in terms of IAQ and thermal comfort. Mitigating and/or forgiveness factors include the presence of adaptive opportunities, such as opening windows, flexibility of indoor spaces and/or control features. Similarly, adequate maintenance of the MVHR system is crucial in ensuring overall system performance.

It is recommended therefore that the Passivhaus Standard should not be adopted in isolation, as overall performance requires a fundamental understanding of the dynamic relationship between the building, the occupant and climate. Moreover, performance in practice requires a certain degree of 'control' over factors, which in reality is difficult to achieve, particularly in social housing. It is suggested therefore that greater attention should be placed on the provision of mitigating or forgiveness factors, and how these may be adopted to provide comfortable and healthy indoor environments, while maintaining optimal energy performance.

11.6 Conclusions

This study aimed to investigate the IAQ and thermal comfort in Passivhaus social dwellings through a UK case study. The findings suggest both measured and perceived IAQ problems, including issues with the perception of thermal comfort and overheating in the homes. A number of issues were identified relating to the use and maintenance of MVHR systems, including lack of knowledge from the building occupants. The findings cannot provide a generalisation of all UK Passivhaus social dwellings, since the number of homes investigated was significantly limited. A further limitation is the relatively small measurement period of 2 days during both summer and winter. However, it does provide interesting insights into IAQ and thermal comfort in these homes, including potential effects of occupant behaviour and activities on IAQ.

Further research is required to investigate the effects of energy efficient design strategies including the Passivhaus standard on IAQ and thermal comfort; to insure occupant health and wellbeing is not sacrificed in the drive towards the reduction of carbon dioxide emissions. Furthermore, the risk of potentially increasing demand for air-conditioning devices in low energy dwellings needs to be addressed to ensure energy savings from reduced heating demand are not off-set by increased demand for cooling. A re-evaluation of energy efficient design strategies may be required to account for future climate predictions and IAQ needs.

Acknowledgements The authors would like to thank Dr. Robin Patrick from ASEP (School of Chemistry, Queen's University Belfast) for help with the measurement and analysis of VOC's and the Housing Association and their tenants for providing access to the case study dwellings.

References

1. HM Government.: Climate Change Act, Chapter 27. 2008:1–103
2. Department for Communities and Local Government: Building a Greener Future: Towards Zero Carbon Development. DCLG publications, West Yorkshire (2006)
3. Zero carbon hub, NHBC.: Zero carbon strategies for tomorrow's new homes. Zero carbon hub & NHBC foundation, Milton Keynes (2013)
4. Cowan, A., Jones, M.R., Forster, W., Heal, A.: Design of Low/Zero Carbon Buildings: Dwelling. Design Research Unit Wales, Cardiff University, Cardiff, pp. 1–45 (2011)
5. Board, The Green Construction: Low Carbon Routemap for the UK Built Environment, pp. 1–93. WRAP, England (2013)
6. Hodgson, G.: An introduction to PassivHaus: A guide for UK application, Information Paper, Watford, BRE Press. 2008;IP 12/08
7. Pelly, H., Hartman, H.: Green sky thinking: Why choose Passivhaus? Architects J. 2013; 2014
8. Cutland, N.: Lessons from Germany's Passivhaus Experience: Informing the Debate. IHS BRE Press/NHBC Foundation, Milton Keynes; F 47:1–26 (2012)
9. Feist, W.: Das Passivhaus, pp. 1–13. Passivhaus Institute, Germany (1999)
10. European Commission. Action Plan for Energy Efficiency: Realising the Potential, Brussels: Commission of the European Communities. 2008; SEC(2006)1173–1175:1–25
11. McLeod, R.S., Hopfe, C.J., Rezgui, Y.: An investigation into recent proposals for a revised definition of zero carbon homes in the UK. Energy Policy **46**, 25–35 (2012)
12. Sullivan, L., Smith, N., Adams, D., Andrews, I., Aston, W., Bromley, K., et al.: Mechanical Ventilation with Heat Recovery in New Homes. Zero Carbon Hub and NHBC, Milton Keynes (2012)
13. Femenias P.: Demonstration projects for sustainable building: Towards a Strategy for Sustainable Development in the Building Sector based on Swedish and Dutch Experience, Ph. D. Thesis, Department of Built Environment & Sustainable Development, Chalmers University of technology, Goteborg, Sweden (2004)
14. Sullivan, L., Smith, N., Adams, D., Andrews, I., Aston, W., Bromley, K., et al.: Mechanical Ventilation with Heat Recovery in New Homes. NHBC, Zero Carbon Hub, London (2013)
15. Sharpe, T., Porteous, C., Foster, J., Shearer, D.: An assessment of environmental conditions in bedrooms of contemporary low energy houses in Scotland. Indoor Built Environ. **23**(3), 393–416 (2014)
16. Howieson, S., Sharpe, T., Farren, P.: Building tight–ventilating right? How are new air tightness standards affecting indoor air quality in dwellings? Build. Serv. Eng. Res. Technol. **35**(5), 475–487 (2013)
17. Sharpe, T., Morgan, C.: Towards low carbon homes Measured performance of four Passivhaus projects in Scotland. Eurosun 2014, 16–19th September 2014, Aix-lesbains, France (2014)
18. Mlecnik, E., Schütze, T., Jansen, S.J.T., De Vries, G., Visscher, H.J., Van Hal, A.: End-user experiences in nearly zero-energy houses. Energy Build. **49**, 471–478 (2012)
19. McLeod, R.S., Hopfe, C.J., Kwan, A., McLeod, R.S., Hopfe, C.J., Kwan, A.: An investigation into future performance and overheating risks in Passivhaus dwellings. Build. Environ. **70**, 189–209 (2013)
20. Hens, H.: Passive houses: What may happen when energy efficiency becomes the only paradigm? ASHRAE Trans. **118**, 1077–1085 (2012)

21. Hasselaar, E.: Health risk associated with passive houses: an exploration. In: Proceedings of Indoor Air (2008)
22. Schnieders J. CEPHEUS–measurement results from more than 100 dwelling units in passive houses, ECEEE 2003 Summer Study Time to turn down the energy demand, Darmstadt: Passive House Institute. European Council for an Energy Efficient Economy: Summer Study 2003:341–51
23. Schnieders, J., Hermelink, A.: CEPHEUS results: measurements and occupants' satisfaction provide evidence for passive houses being an option for sustainable building. Energy Policy 34, 151–171 (2006)
24. Feist, W., Schnieders, J., Dorer, V., Haas, A.: Re-inventing air heating: Convenient and comfortable within the frame of the passive house concept. Energy Build. 37, 1186–1203 (2005)
25. Mlecnik, E., Schütze, T., Jansen, S., de Vries, G., Visscher, H., van Hal, A.: End-user experiences in nearly zero-energy houses. Energy Build. 49, 471–478 (2012)
26. Schiano-Phan, R., Ford, B., Gillott, M., Rodrigues, L.: The passivhaus standard in the UK: Is it desirable? Is it achievable. PLEA 2008—Towards Zero Energy Building: 25th PLEA International Conference on Passive and Low Energy Architecture, Conference Proceedings, Dublin 22–24th Oct 2008
27. McGill, G., Oyedele, L.O., Keeffe, G., Keig, P.: Indoor air quality and the suitability of mechanical ventilation with heat recovery (MVHR) systems in energy efficient social housing projects: perceptions of UK building professionals. Int. J. Sustain. Build. Technol. Urban Dev. 5, 240–249 (2014)
28. McGill, G., Oyedele, L., Keeffe, G.: Indoor air-quality investigation in code for sustainable homes and Passivhaus Dwellings: A case study. World J. Sci. Technol. Sustain. Dev. 12 (2015)
29. Raw, G., Whitehead, C., Robertson, A., Burge, S., Kely, C., Leinster, P.: A Questionnaire for Studies of Sick Building Syndrome: A Report to the Royal Society of Health Advisory Group on Sick Building Syndrome. Building Research Establishment, Watford (1995)
30. Raw, G.J., Roys, M.S., Whitehead, C., Tong, D.: Questionnaire design for sick building syndrome: An empirical comparison of options. Environ. Int. 22, 61–72 (1996)
31. Porteous, C.D.: Sensing a Historic Low-CO_2 Future. Chemistry, Emission Control, Radioactive Pollution and Indoor Air Quality. Intech, Rijeka, Croatia (2011)
32. Appleby, A.: Indoor air quality and ventilation requirements. In: Curwell, S., March, C., Venables, R. (eds.) Buildings and Health, the Rosehaugh Guide to the Design, Construction, Use and Management of Buildings, pp. 167–193. RIBA Publications, London (1990)
33. Seppänen, O., Fisk, W., Mendell, M.: Association of ventilation rates and CO2 concentrations with health and other responses in commercial and institutional buildings. Indoor Air 9, 226–252 (1999)
34. Wargocki, P.: The effect of ventilation in homes on health. Int. J. Vent. 12, 101–118 (2013)
35. Environmental Protection Agency (EPA).: National Institute for Occupational Safety and Health (NIOSH). Building Air Quality—A Guide for Building Owners and Facility Managers, pp. 91–114. DHHS (NIOSH), Washington DC (1991)
36. CIBSE.: Climate Change and the Indoor Environment: Impacts and Adaptation. Chartered Institution of Building Services Engineers, London (2005)
37. Boer, R.D., Kuller, K.: Mattresses as a winter refuge for house-dust mite populations. Allergy 52, 299–305 (1997)
38. Arlian, L.G.: Water balance and humidity requirements of house dust mites. Exp. Appl. Acarol. 16, 15–35 (1992)
39. Korsgaard, J.: House-dust mites and asthma. A review on house-dust mites as a domestic risk factor for mite asthma. Allergy 53, 77–83 (1998)
40. Harving, H., Korsgaard, J., Dahl, R.: House-dust mites and associated environmental conditions in Danish homes. Allergy 48, 106–109 (1993)

41. Platts-Mills, T.A., de Weck, A.L., Aalberse, R., Bessot, J., Bjorksten, B., Bischoff, E., et al.: Dust mite allergens and asthma—a worldwide problem. J. Allergy Clin. Immunol. **83**, 416–427 (1989)

42. Korsgaard, J.: Epidemiology of house-dust mites. Allergy **53**, 36–40 (1998)

43. Ormandy, D., Ezratty, V., Ormandy, D., Ezratty, V.: Health and thermal comfort: From WHO guidance to housing strategies. Energy Policy **49**, 116–121 (2012)

44. World Health Organisation: Health impact of low indoor temperatures. Report on a WHO meeting **11–14**(1987), 1–27 (1985)

45. World Health Organization: Air Quality Guidelines for Europe, 2nd edn. Regional Office for Europe, WHO, Copenhagen (2000)

46. Coward, S.K.D., Brown, V.M, Crump, D.R, Raw, G.J, Llewellyn, J.W.: Indoor Air Quality in Homes in England: Volatile Organic Compounds; BR 445:1–64. IHS BRE Press, Watford (2002)

47. Coward, S., Llewellyn, J., Raw, G., Brown, V., Crump, D., Ross, D.: Indoor Air Quality in Homes in England. BRE Press and IHS, Watford (2001)

48. German Working Group on Indoor Guideline Values of the Federal Environmental Agency and the State's Health Authority.: Health evaluation of carbon dioxide in indoor air | Gesundheitliche Bewertung von Kohlendioxid in der Innenraumluft. Bundesgesundheitsblatt - Gesundheitsforschung - Gesundheitsschutz 2008; 51:1358–69

49. Dengel, A., Swainson, M.: Overheating in New Homes: A Review of the Evidence; NF46:1–40. NHBC and the Zero Carbon Hub, Milton Keynes (2012)

50. Larsen, T.S.: Overheating and insufficient heating problems in low energy houses up to now call for improvements in future. REHVA J. **48**, 36–40 (2011)

51. Rodrigues, L.T., Gillott, M., Tetlow, D.: Summer overheating potential in a lowenergy steel frame house in future climate scenarios. Sustain. Cities Soc. **7**, 1–15 (2013)

52. McCready, D., Arnold, S.M., Fontaine, D.D.: Evaluation of potential exposure to formaldehyde air emissions from a washing machine using the IAQX model. Hum. Ecol. Risk Assess. **18**, 832–854 (2012)

53. Steinemann, A.C.: Fragranced consumer products and undisclosed ingredients. Environ. Impact Assess. Rev. **29**, 32–38 (2009)

54. Steinemann, A.C., MacGregor, I.C., Gordon, S.M., Gallagher, L.G., Davis, A.L., Ribeiro, D. S., et al.: Fragranced consumer products: Chemicals emitted, ingredients unlisted. Environ. Impact Assess. Rev. **31**, 328–333 (2011)

Printed in the United States
By Bookmasters